Communications in Computer and Information Science 2399

Rationale

The CCIS series is devoted to the publication of proceedings of computer science conferences. Its aim is to efficiently disseminate original research results in informatics in printed and electronic form. While the focus is on publication of peer-reviewed full papers presenting mature work, inclusion of reviewed short papers reporting on work in progress is welcome, too. Besides globally relevant meetings with internationally representative program committees guaranteeing a strict peer-reviewing and paper selection process, conferences run by societies or of high regional or national relevance are also considered for publication.

Topics

The topical scope of CCIS spans the entire spectrum of informatics ranging from foundational topics in the theory of computing to information and communications science and technology and a broad variety of interdisciplinary application fields.

Information for Volume Editors and Authors

Publication in CCIS is free of charge. No royalties are paid, however, we offer registered conference participants temporary free access to the online version of the conference proceedings on SpringerLink (http://link.springer.com) by means of an http referrer from the conference website and/or a number of complimentary printed copies, as specified in the official acceptance email of the event.

CCIS proceedings can be published in time for distribution at conferences or as post-proceedings, and delivered in the form of printed books and/or electronically as USBs and/or e-content licenses for accessing proceedings at SpringerLink. Furthermore, CCIS proceedings are included in the CCIS electronic book series hosted in the SpringerLink digital library at http://link.springer.com/bookseries/7899. Conferences publishing in CCIS are allowed to use Online Conference Service (OCS) for managing the whole proceedings lifecycle (from submission and reviewing to preparing for publication) free of charge.

Publication process

The language of publication is exclusively English. Authors publishing in CCIS have to sign the Springer CCIS copyright transfer form, however, they are free to use their material published in CCIS for substantially changed, more elaborate subsequent publications elsewhere. For the preparation of the camera-ready papers/files, authors have to strictly adhere to the Springer CCIS Authors' Instructions and are strongly encouraged to use the CCIS LaTeX style files or templates.

Abstracting/Indexing

CCIS is abstracted/indexed in DBLP, Google Scholar, EI-Compendex, Mathematical Reviews, SCImago, Scopus. CCIS volumes are also submitted for the inclusion in ISI Proceedings.

How to start

To start the evaluation of your proposal for inclusion in the CCIS series, please send an e-mail to ccis@springer.com

Ashish Bagwari · Jorge Luis Victoria Barbosa ·
Eduard Babulak · Durg Singh Chauhan
Editors

Emerging Wireless Technologies and Sciences

Second International Conference, ICEWTS 2024
Uttarakhand, India, October 6–7, 2024
Proceedings

Springer

Editors
Ashish Bagwari
WIT, VMSBUTU
Dehradun, Uttarakhand, India

Eduard Babulak
University of Maryland
Baltimore, MD, USA

Jorge Luis Victoria Barbosa
University of Vale do Rio dos Sinos
São Leopoldo, Brazil

Durg Singh Chauhan
GLA University
Mathura, Uttar Pradesh, India

ISSN 1865-0929 ISSN 1865-0937 (electronic)
Communications in Computer and Information Science
ISBN 978-3-031-87885-5 ISBN 978-3-031-87886-2 (eBook)
https://doi.org/10.1007/978-3-031-87886-2

This Springer imprint is published by the registered company Springer Nature Switzerland AG
The registered company address is: Gewerbestrasse 11, 6330 Cham, Switzerland

Preface

On behalf of the **Advanced And Innovative Research Laboratory (AAIR Lab) Dehradun, India**, we are pleased to welcome all the participants of the **Second International Conference on Emerging Wireless Technologies and Sciences (ICEWTS 2024), held on 6th–7th October, 2024**. The ICEWTS 2024 provided a prominent international forum for researchers and practitioners to exchange information regarding novel aspects of technology, application, and service development within the multidisciplinary framework of communication engineering, wireless sensor networks, electrical engineering, renewable energies, materials science, physics, mathematics, computer science, environment, and many more.

The Keynote Speakers, Distinguished Lecturers, Invited Talks, and paper presenters at the ICEWTS 2024 conference concentrated on current technologies in the areas of wireless, networks, and science, and their issues & challenges.

This year, the response to the conference's call for papers was outstanding. Many individuals and organizations contributed to the success of this conference. Together with the Technical Program Committee, they worked diligently to select papers and speakers that met the criteria of high quality and relevance to our various fields of interest. We are grateful for all their hard work and efforts. It takes time and effort to review a paper carefully, and every member of the Technical Program Committee is to be commended for his or her contribution to the success of this conference. The TPC Chairs, Co-Chairs, and members had the unenviable task of coordinating the peer review process and putting together an outstanding technical program. They deserve a lot of thanks. The peer review process has been done by three double-blinded reviewers per paper who reviewed the authors' papers, and based on their recommendation ICEWTS 2024 released the paper acceptance/rejection email to the concerned authors. These conference proceedings of ICEWTS 2024 contain 4 full and 4 short papers from the 430 submissions. The success of this conference would not have been possible without the dedication and efforts of the members of the Advisory Committee, Publicity Committee, Steering Committee, and other committees.

We would like to extend our gratitude to the team members of Springer's Communications in Computer and Information Science (CCIS) and to the Technical Session Chairs for their all-out support and encouragement.

It was an honor and a pleasure for us to accept the responsibilities and challenges of serving as Conference General Chairs. We sincerely hope that everyone who attended the conference enjoyed themselves and learned something new. Also, we appreciate and thank all those researchers who provided support in various ways. See you again in 2025 at the next version of this conference.

For details on upcoming conferences, journals, webinars, and grant-writing services, kindly visit our website (www.icewts.aairlab.com or www.aairlab.com).

February 2025

Ashish Bagwari
Jorge Luis Victoria Barbosa
Eduard Babulak
Durg Singh Chauhan

Organization

General Chairs

Ashish Bagwari	WIT, VMSBUTU, India
Eduard Babulak	National Science Foundation, USA & University of Maryland University College, USA
Jorge Barbosa	University of Vale do Rio dos Sinos, Brazil
D. S. Chauhan	GLA Mathura, India

Program Committee Chairs

K. S. Sastry	Namibia University of Science and Technology, Namibia
Ivan Perl	ITMO University, Russia
Ashraf Samarah	University of Cincinnati, USA
Jyotshana Bagwari	AAIR Lab, India
Rahul Tiwari	AAIR Lab, India

Steering Committee

Dwarkadas Pralhaddas Kothari	IIT Delhi, India
A. S. Vidyarthi	Bikaner Technical University, India
Gauri Mittal	University of Guelph, Canada
Narendra S. Chaudhari	Visvesvaraya National Institute of Technology Nagpur & IIT Indore, India
Ruidan Su	Dalian University of Technology, China
Alhad Kuwadekar	University of South Wales, UK
Nalin Asanaka	La Trobe University, Australia
Gagangeet Aujla	Durham University, UK
Prosanta Gope	University of Sheffield, UK
Brahmjit Singh	National Institute of Technology, Kurukshetra, India
Vaibhav Bhatnagar	STE, Bowen Orsay, France
Sunil K. Somani	Medi-Caps University, India
Manisha Khulbe	NSUT East Campus, India
Zakia Hammouch Errachidia	Moulay Ismail University, Morocco

Ganesh Nagrajan	Pondicherry Engineering College, India
Ali L. Al-Mosawi	Hilla, Babylon, Iraq
Haider Alrikabi	Wasit University, Iraq
A. Jayanthiladevi	Srinivas University, India
Teguh Sugiarto	Universitas Budi Luhur, Indonesia
Shaymaa Abbas Abdulsada	University of Miskolc, Hungary
Siamak Hosseinzadeh	Islamic Azad University, Iran
Jiban Shrestha	Nepal Agricultural Research Council, Nepal
M. Mahbubur Rahman	Jahangirnagar University, Bangladesh

Program Committee

Mohit Mittal	National Institute for Research in Computer Science and Automation (INRIA), France
Umit Cali	Norwegian University of Science and Technology, Norway
Rajendra Choudhary	Leibniz University Hannover, Germany
Saurabh Khandelwal	Oxford Brookes University, UK
Zeeshan Pervez	University of the West of Scotland, UK
Juned Ahmed Siddiqui	Medi-Caps University, India
Bhola Jha	GBPIET, India
Vandana Roy	GGITS, India
Krishan Kumar	National Institute of Technology, Uttarakhand, India
Ashish Ranjan	National Institute of Technology, Manipur, India
Surendra Singh	National Institute of Technology, Uttarakhand, India
Hemant Kumar Singhal	National Institute of Technology, Uttarakhand, India
Ruchi Sharma	Bharati Vidyapeeth College of Engineering, India
Ashish Gupta	Dr. Shyama Prasad Mukherjee University, India
Pradeep Kundu	University of Strathclyde, UK
Udayabalan Balasingam	East Point College of Engineering and Technology, India
Vivek Kushwah	CBIT (Autonomous), Affiliated to Osmania University, India
Sachin Puntambekar	Medi-Caps University, India
Shikar Gupta	Indian Institute of Technology, Roorkee, India
Narinder Sharma	Amritsar College of Engineering and Technology, India
Mayank Swarnkar	Indian Institute of Technology BHU, Varanasi, India

Deepak Jain	Chongqing University of Posts and Telecommunications, China
Meena Jha	Central Queensland University, Australia
Pokkuluri Kiran Sree	Shri Vishnu Engineering College for Women, India
Indhumathi K.	Saveetha Engineering College, India
Mukesh Kumar	National Institute of Technology Kurukshetra, India
Ripal Ranpara	Atmiya University, India
S. Kannadhasan	Cheran College of Engineering, India
Nitin Arora	Indian Institute of Technology Roorkee, India
Praphula Kumar Jain	MITS Gwalior, India
Manikandan K.	Vellore Institute of Technology, Vellore, India
V. M. Thakkar	GBPIET, India
Varun Mishra	Amity University, Gwalior, India
Rajiv Verma	Indian Institute of Information Technology, Sonepat, India
Richa Gupta	MSIT, India
Rajesh Kumar	University of Electronic Science and Technology of China, China
Bharti Nagpal	Netaji Subhas University of Technology (East Campus), India
Neeraj Kumar	Thapar Institute of Engineering and Technology, India
Vikash Singh	Manipal Institute of Technology, India
Sayed Sadaf Ali	École Nationale Supérieure de L'électronique et de ses Applications (Ensea), France

Additional Reviewers

Olga Perl	ITMO University, Russia
Madhusudan Sing	Woosong University, South Korea
Ihsan Mohd Yassin	Universiti Teknologi Mara, Malaysia
Yubo Chen	Tsinghua University, China
Mithun Mukherjee	Guangdong University of Petrochemical Technology, China
Lili He	San José State University, USA
Alberto H. F. Laender	Universidade Federal de Minas Gerais, Brazil
Gautam Srivastava	Brandon University, Canada
Ning Zhang	Washington University in St. Louis, USA
Chia Y. Han	University of Cincinnati, USA
Sourav Mazumder	Bmcloud and Cognitive Lab, USA

Harald Gjermundrød	University of Nicosia, Cyprus
Evaggelos Spyrou	University of Thessaly, Greece
Sándor Szénási	Óbuda University, Hungary
Matevž Pustišek	University of Ljubljana, Slovenia
Zoran Bojkovic	Engineering Academy of Serbia, Serbia
Parag Chatterjee	University of the Republic, Uruguay
Hakan Koyuncu	Altinbas University, Turkey
Abul Bashar	Prince Mohammad Bin Fahd University, Saudi Arabia
Mohammad Kamrul Hasan	Universiti Kebangsaan Malaysia/National University of Malaysia, Malaysia
Saurabh Yadav	Yeungnam University, South Korea
Mohammad Talha	King Fahd University, Saudi Arabia
Hamid Ali Abed Al-Asadi	Iraq University, Iraq
Siamak Hoseinzadeh	Municipality of Sari, Iran
Ahmed Nabih Zaki Rashed	Menoufiya University, Egypt

Contents

Sunflower-Inspired Precision: Automatic Solar Tracking System for Enhanced Energy Harvesting

Juned A. Siddiqui(✉), Lavesh Kushwah, Nikhil Asati, Deepak Yaduwanshi, and Ranjeet Kushwah

Department of Electronics, Medi-Caps University Indore, Indore, India
{juned.siddiqui,en20el301055,en20el301065,en20el301029, en20el301074}@medicaps.ac.in

Abstract. The paper investigates the feasibility and the effectiveness of a sunflower based heliotropic mechanism for tracking solar PV panels, aiming to optimize the efficiency of solar energy systems. Inspired by the natural heliotropism observed in sunflowers, the study explores design and the implementation of a nature-inspired solar photovoltaic panel tracking system. The research examines the potential of utilizing pulvinus, a joint-like thickening at the base of sunflowers responsible for heliotropism, to develop a sustainable and low-maintenance tracking mechanism. The study have shown a detailed analysis about the sunflower-based Solar tracking System, discussing its design principles, mechanism, and potential applications in optimizing the efficiency of solar PV systems. Additionally, the paper explores the integration of sunflower-inspired tracking mechanisms with bio-substances for realistic biotechnological design. The effectiveness and commercial viability of proposed methodology have been demonstrated by experimental results, highlighting its ability to maximize solar energy capture under various weather conditions. Furthermore, the paper addresses the energy crisis in developing nations like India and emphasizes the importance of renewable energy, particularly solar energy, in addressing this challenge. The study evaluates the effectiveness of an Arduino based Automatic sunflower solar tracking system compared to static solar panels. Through hardware and software implementation, the research demonstrates higher performance of system in terms of the fundamental quantities such as power output, voltage and current with the solar tracking system. Specifically, the results show that automatic sunflower solar tracking system produces an additional 0.72 watts of average power compared to fixed or single-axis systems. In conclusion, the research underscores the potential of sunflower-inspired solar tracking systems as a robust, sustainable, and economically viable solution for optimizing the efficiency of solar PV energy conversion. The findings contribute to advancing renewable energy technologies and offer insights into maximizing solar energy harvest for various applications, particularly in regions facing energy scarcity.

Keywords: Photovoltaic tracking · Automated solar tracker · Energy harvesting system · Weather resistance · Control loop · Tracking angle adjustment · Power consumption optimization

A. Bagwari et al. (Eds.): ICEWTS 2024, CCIS 2399, pp. 1–7, 2025.
https://doi.org/10.1007/978-3-031-87886-2_1

1 Introduction

Solar energy, a prominent form of renewable energy, has gained widespread popularity due to its clean and abundant nature. Its utilization spans across various sectors including households, industries, and educational institutions. While fixed solar panels have been traditionally used, recent advancements have highlighted the superior efficiency of movable solar tracking systems. These systems are equipped with light based sensors, ensuring optimal alignments of all solar panels with the position of the sun throughout the day, maximizing energy capture. In countries like India, where solar energy potential remains largely untapped, research efforts are directed towards enhancing solar panel effectiveness. The focus lies on innovative technologies such as dual-axis solar trackers, which promise greater precision and efficiency. By leveraging phototropism principles observed in plants, researchers aim to develop robust tracking mechanisms for improved solar energy utilization. This pursuit aligns with the broader goal of enhancing solar energy efficiency to meet growing energy demands sustainably.

2 Literature Summary

Several studies have explored the design and implementation [1, 2] of automatic solar tracking systems using Arduino microcontrollers, contributing valuable insights to the field of renewable energy. Author [8–10] developed a dual axis solar tracking system that demonstrated the superior performance compared to static solar panels, showing increased voltage, current, and power output. Similarly, Authors [5, 6] focused on an Arduino based automatic solar tracking system incorporating light dependent resistors(LDRs) and servo motors to maximize the solar energy capture.

Dhanabal, R., V. Bharathi [6] presented an Arduino-based sun tracking solar panel system, highlighting the use of sensors and actuators for optimal sunlight capture. Authors [3] introduced an Arduino-based automatic sun tracking system, employing light sensors and servo motors for solar panel orientation adjustment. Additionally, Anyaka, B. et al. [7] discussed the development of an automatic solar tracking system utilizing Arduino microcontrollers, offering insights into hardware and software components.

The global energy crisis has underscored the importance of alternative energy sources, such as solar power, in meeting increasing energy demands sustainably. Solar energy, as an abundant and eco-friendly resource, has garnered significant attention for its potential to address energy scarcity. Advances in technology have facilitated the utilization of solar energy for various applications like electricity generation through photovoltaic (PV) systems.

.Solar tracking devices, which orient photovoltaic arrays to maximize sunlight exposure, played a crucial role enhancing the efficiency of solar energy system. These devices utilize microprocessors or feedback controllers show that they can determine the position of moving sun and can adjust the orientation of the used solar panels accordingly. Active solar tracking systems, such as those employing light-detecting resistors (LDRs) or ambient light sensors, continuously monitor sunlight intensity and adjust panel orientation to optimize energy capture.

However, active solar tracking systems may face challenges such as inefficiency during cloudy weather conditions and complexity in positioning mechanisms. The constant movement of panels driven by motors can lead to energy consumption and decreased system efficiency.

3 Methodology

The automatic solar tracker using Arduino project's methodology initiates with the Arduino microcontroller assessing ambient light levels via light-dependent resistors (LDRs). Once light exceeds a predetermined threshold, signifying sunlight availability, the system activates. An N20 gear motor is then engaged, rotating anticlockwise through the L293D Driver until a limit switch signals the solar panel's complete extension.

Upon sunlight detection, data from additional LDRs are transmitted to the Arduino, enabling determination of the sun's position. Servo motors, with a rotational capacity of up to 170 degrees, adjust the panel's orientation accordingly to ensure optimal alignment of the solar panels with the trajectory of sun throughout the day.

As daylight wanes, a fifth LDR detects diminishing light levels, prompting the clockwise rotation of the N20 motor. This movement continues until a second limit switch is activated, returning the panel to its initial position, or "home position."

The process repeats cyclically, activating the system daily at sunrise to track the movement of the sun across the sky, optimizing solar panel orientation for maximal energy capture. Custom Arduino hardware, including shields and sensor holders, is utilized alongside tailored code to execute the system's functions efficiently.

Servo motors are calibrated to maintain the panel within a specified range of motion, while LDRs provide real-time feedback on sunlight intensity, guiding precise adjustments to panel orientation. This iterative process ensures that the solar panel continually adapts to the sun's position, maximizing energy harvesting potential.

Overall, this methodology facilitates effective automatic solar tracking, enhancing solar energy utilization and system efficiency for sustainable energy generation.

4 Block Diagram and Simulated Model

See Figs. 1 and 2.

5 Results, Analysis and Discussion

5.1 Efficiency of the Automatic Sunflower Solar Tracking System

A better output efficiency of the automatic solar tracking system can lead us to the success of our project. Now for determining the desired output, we can calculate the efficiency of the automatic solar tracking system. Now to calculate the efficiency, multiply the amount of the sunlight that is hitting the earth's surface in given area.

Now this equation is used to satisfy the condition to get efficiency of our system:

$$\varepsilon = (A - B) - C > 0 \tag{1}$$

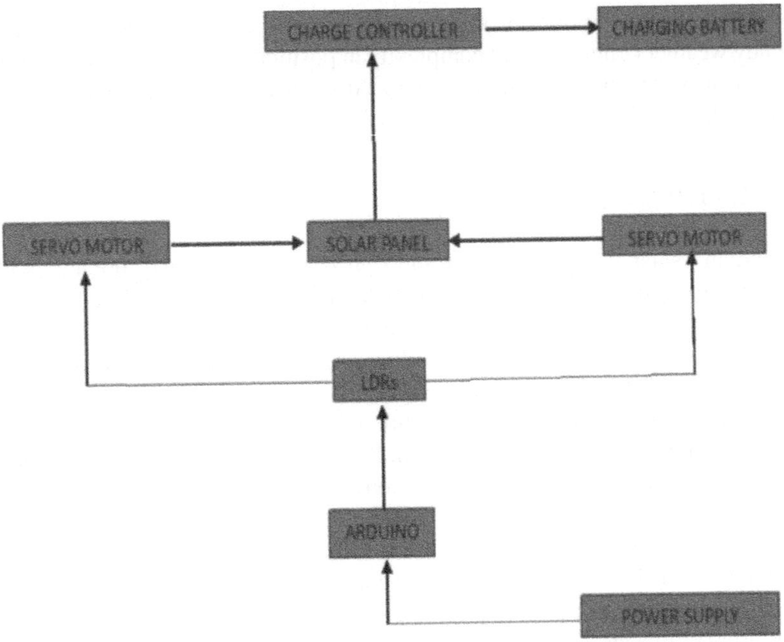

Fig. 1. Bock diagram of Automatic solar tracking system.

Fig. 2. Simulated model

where

A = the electric energy which is produced by the solar panels with tracking system.

EF = the electric energy which is produced by the same solar panels without tracking system or when fixed.

EC = the energy consumed for orienting the solar panels Performance of Static Solar Panels (without Tracking system).

5.2 Performance

Table 1. Performance of Static Solar Panels (without Tracking system)

Time (Hrs)	Current (A)	Voltage (V)	Power (W)
8am	0.09	5.0	0.45
9am	0.15	6.5	0.98
10am	0.25	11	2.75
11am	0.28	11.4	3.19
12pm	0.32	12.3	3.94
1pm	0.34	12.8	4.35
2pm	0.29	11.6	3.36
3pm	0.26	11.0	2.86
4pm	0.16	8.4	1.34
5pm	0.11	6.2	0.68

Table 2. Performance of Solar Panels with Tracking system

Time (Hrs)	Current (A)	Voltage (V)	Power (W)
8am	0.24	11.7	2.80
9am	0.28	12.2	3.42
10am	0.26	12.1	3.15
11am	0.31	12.6	3.91
12pm	0.34	12.8	4.35
1pm	0.30	13.0	3.9
2pm	0.29	12.7	3.68
3pm	0.26	11.3	2.93
4pm	0.20	10.2	2.04
5pm	0.14	6.6	0.92

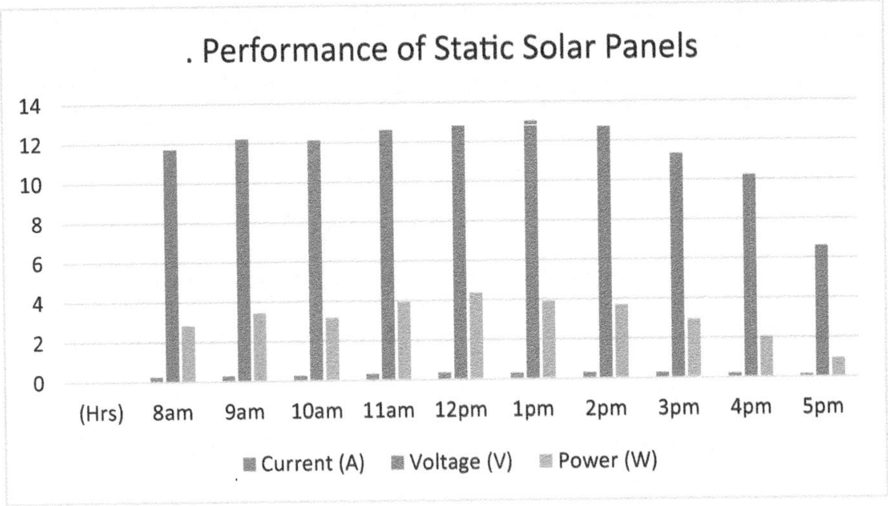

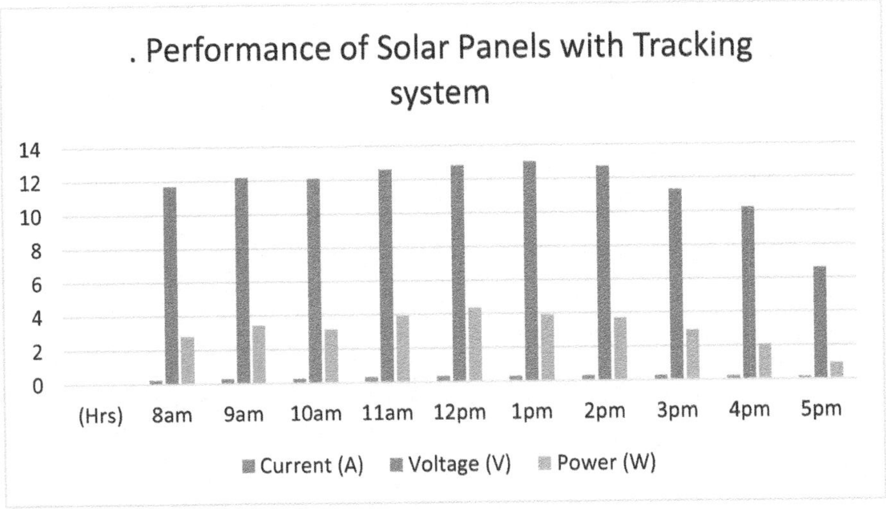

From the above data, we can conclude that we have got average power of 2.39 watts from solar panel without tracking and 3.11watts of average power obtained from solar panel with tracking system present. Even neglecting the power consumption of the motor we got 30.12% as a inproved efficiency (Tables 1 and 2).

In our project, the hardware design and implementation of an automatic solar tracking solar panel system was proposed. Our findings showed that the solar tracking system increased the efficiency of the solar panels. The solar tracking solar panel system is completely automatic and ensures minimum cost. Thus, it is a dual axis system that rotates 360 degree that maximizes efficiency. Normally, a solar panel can converts only 30–40% of the solar radiation incident into the electrical energy. An automated system is required to achieve a system generate a constant output, capable of continuously rotating the solar panel in the position of sun. The sun tracking system was developed as a prototype to solve the problem. It was automatic system and kept the solar panel facing forward towards the sun as long as sunlight was visible.

6 Conclusion

The aimed of this work to enhance solar panel efficiency by creating a system that mimics sunflowers' ability to face the sun. By carefully planning and applying engineering skills, we developed an affordable solar tracking system. Our goal was to combat global warming by increasing the use of clean energy sources like solar power. Our system can boost solar panel electricity production by 5–8%, a significant improvement for areas lacking reliable electricity access. Its unique design allows it to track the sun in two directions, optimizing energy capture throughout the day. Similar to sunflowers, our system adjusts itself to follow the sun's path, ensuring maximum sunlight exposure. This adaptability makes it ideal for remote villages, offering a reliable electricity source without relying on distant power plants. By generating electricity locally, our system increases reliability and simplifies maintenance, contributing to sustainable energy solutions in underserved communities.

References

1. Batayneh W, Owais A, Nairoukh M (2013) An intelligent fuzzy based tracking controller for a dual-axis solar PV system. Autom. Construct., **29** (2013)
2. Gupta, B., Sonkar, N., Bhalavi, B.S., Pankaj Edla, J.: Design, construction and effectiveness analysis of hybrid automatic solar tracking system for amorphous and crystalline solar cells. Am. J. Eng. Res. **2**(10), 221–228 (2013)
3. Akhil K (March 25, 2016) Solar tracking system. Retrieved on October 3rd, 2019 from: https://www.slideshare.net/Reejasunil/solar-tracking-system-60035017
4. Gajadharane MR, Gijare NA, Joshi MH (June, 2016) Design and development of mechanical Solar Tracking System. Retrieved on October 7, 2019 from https://www.academia.edu/26526209/Design_and_Development_of_Mechanical_Sol ar_Tracking_System
5. Mishra J, Thakur R, Deep A (May, 2017) Arduino based dual axis smart solar tracker. Retrieved on September 29, 2019 from https://media.neliti.com/media/publications/239849-arduino-based-dual-axis-smartsolar-trac-8ab31d02.pdf
6. Dhanabal, R., Bharathi, V., Ranjitha, R., Ponni, A., Deepthi, S., Mageshkannan, P.: Comparison of efficiencies of solar tracker systems with static panel single axis tracking system and dual axis tracking system with fixed mount. Int. J. Eng. Technol. (IJET) **5**(2), 1925–1933 (2013)
7. Anyaka BO, Ahiabuike DC, Mbunwe MJ (2013) Improvement of PV systems power output using sun-tracking techniques. Int. J. Comput. Eng. Res., **3**(9):80–98
8. Pradeep KPJ, Sai Prasad Reddy K, Chandra Mouli C, Nagabhushan Raju K (2014) Development of dual-axis solar tracking using arduino with lab VIEW. Int. J. Eng. Trends Technol. (IJETT), **17** (2014)
9. Deepthi S, Ponni A, Ranjitha R, Dhanabal R (2013) Electronics and communications engineering, SENSE, VIT UNIVERSITY, Comparison of Efficiencies of Single-Axis Tracking System and Dual-Axis Tracking System with Fixed Mount. Int. J. Eng. Sci. Innov. Technol. (IJESIT), **2**(2)
10. Brahmeswara Rao V, Durga Harish Kumar K, Upendra Kumar NV, Deepak K (2017) Arduino based two axis solar tracking by using servo mechanism. Int. J. Modern Trends Sci. Technol., **03**(02)

Artificial Neural Network Based ECG Feature Extraction Using Wavelet Transform

Priyanka Agrawal[1], Vanya Arun[1], and Amlan Basu[2(✉)]

[1] Department of Electrical & Electronics Engineering, IILM University, Noida, UP, India
[2] Centre for Ultrasonic Engineering, University of Strathclyde, Glasgow, UK
amlan.basu@strath.ac.uk

Abstract. In this research, Automatic techniques to detect diseases have been developed because of requirements of continuous attention to patient having heart diseases. This research deals with implementation of Artificial neural network methods for analyzing ECG (Electrocardiogram) signals with a focus on early and accurate detection. Feature extraction of ECG signal plays vital role in cardiovascular diseases. ECG signal is decomposed using wavelet transform and then feature extracted of decomposed ECG signal are given as input to Neural Network. The wavelets used for decomposition are Daubechies and Symmetric. The selection of detail coefficient d4 had been done based on the following important parameters i.e. Energy, Frequency and Correlation. The overall of detection using db6 and sym11 were 96.65% and 84.37%. In this work, study of the classification of ECG signal has been done in detail by using computational methods effectively for early cardiovascular diagnosis. Coefficients of discrete wavelet transforms are used for analyzing ECG signals in conjunction with the Artificial Neural network (ANN). Three different types of ECG data have been used normal sinus rhythm, supra ventricular arrhythmia and atrial fibrillation. Decomposition and Classification of ECG signals using discrete wavelet transform and Artificial Neural Network have been successfully designed. The method has been implemented on 18 subjects. The results show that proposed method is effective for classification of normal and cardiac arrhythmia with an overall accuracy of 97.5%

Keywords: ECG · Neural Network · Wavelet · Feature Extraction · Morphological Features · Statistical Features

1 Introduction

1.1 A Subsection Sample

This paper focuses to extract morphology and statistical parameters from the ECG signal using the Discrete Wavelet Transform and Artificial Neural Network technique. ECG signal is decomposed using discrete wavelet transform then features of Decomposed ECG signal are extracted and these features are trained using back propagation neural network.

A. Bagwari et al. (Eds.): ICEWTS 2024, CCIS 2399, pp. 8–22, 2025.
https://doi.org/10.1007/978-3-031-87886-2_2

ECG is a diagnostic methodology that records the biological signals activity of heart from the chest. It's a common and painless test used to quickly detect heart problems and monitor the heart's health. ECG is used to find heart problems and health of heart in very quick way. This is non- invasive and painless test. ECG may be used to find Arrhythmias, coronary artery diseases and heart attack. During the process of diagnosis, many electrodes are used on human limbs [1]. These electrodes in pairs are called lead. Analysis of ECG signal systems have a vital role in assisting medical staff by for fast treatment in hospitals and enhancing quality of life by finding abnormality at an early age. ECG signals have been used by doctors to find diseases like arrhythmia and myocardial infarction An ECG signal has P, QRS complex, and T waves as shown in Fig. 1. The P wave shows atrial depolarization. The QRS complex reflects ventricular depolarization. The T wave shows ventricular repolarization. Cardiac diseases are detected by analyzing these waves. Spikes can be seen in waves due to artifacts. Such artifacts are in waves due to movements in body of patients and electrode and interferences [2]. So, these noise and artifacts should be filtered from ECG signals so that accurate analysis of ECG signal can be done. For removing noise, different time and frequency domain transforms have been used. Previous research involves Discrete Wavelet Transform, Hermitian Basis and other methods but now a days Wavelet transform is most widely used [3].

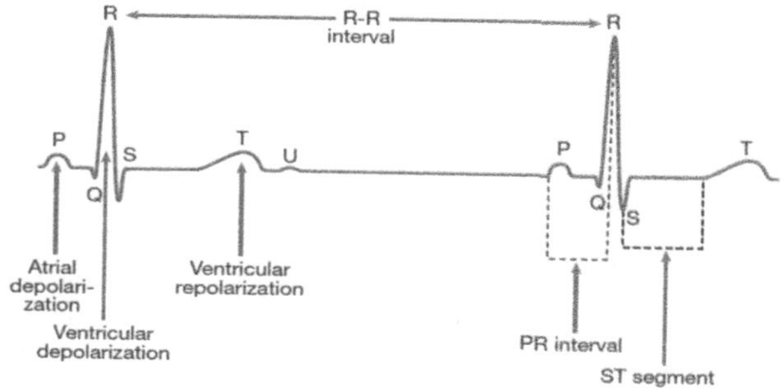

Fig. 1. ECG Waveform.

2 Methodology

Various methodologies of automatic diagnosis have been done. Neural network and wavelet tool box has been used for decomposition of ECG and its classification. Data acquisition has been done from MIT-BIH data base available at physionet.org for performance evaluation. For the analysis of signal various tools are used such as wavelet, neural network. During the analysis of ECG signals, variations in features of ECG signals is the main problem. Such variations occur not only in different patients or group of patients but also with in the same patient. In our present research, different ECG samples of different diseases have been taken.In order to classify ECG signals based on diseases, following ECG signal analysis pipeline will be done:Selections of wavelet, particular detailed coefficient, feature extraction, process of classification using ANN.

2.1 Data Acquisition

It is mandatory that Standard database should be selected to analyze ECG signal. ECG signals used in this paper were taken from Physionet MIT-BIH.ECG signals are denoted by header file and a mat file [4]. Header file has detailed information about sampling frequency, number of ECG leads, and information about patient.ECG signals used in this paper are: MIT BIH Normal, Atrial fibrillation and super ventricular arrhythmia. 6 subjects of Normal ECG, Atrial fibrillation and Supra ventricular is taken, half of them have been used for training and remaining for testing. All classes, sampling frequency of each class, number of samples are shown in Table 1. Data base contains total 18 records of 1 min recording. 1024 samples of each record have been taken in this paper.

Table 1. Data

Classes	Sampling frequency	Gain (mv)
N (Normal)	128 Hz	200
SV (Supra Ventricular Arrhythmia)	128 Hz	200
AF (Atrial Fibrillation)	250 Hz	200

2.2 Wavelet Transform

Wavelet transform deals with processing method for non-stationary signals with enough scale values and shifting in time. The wavelet transform represents signals in different resolutions by dilating and compressing in different functions. The wavelet transform is considered a valuable tool to analyse data from different fields such as biomedical, mathematics, mechanics etc. It has now been applied to a different types of biomedical signals: the ECG, EMG, EEG, image compression. It is very important to select mother wavelet for preprocessing of ECG. There are different basis wavelets for example Symmlet, Harr, Daubechies wavelets etc. The main advantage of wavelet method is the availability of wavelet functions, so making easy to choose the most suitable for the signal under investigation.

There are some predefined criteria to select optimal wavelet transform. Suitable wavelet for ECG signal is Daubechies wavelet. This wavelet families have similarity with QRS complex and energy spectrums of Daubechies wavelets are concentrated around low frequencies[5]. The generated cross correlation coefficient of the ECG signal with various wavelet filters is shown in Fig. 2. Results strengthens the suitability of Daubechies wavelet filter of order 8 is an optimal one for the ECG signal processing. So, in this paper, Daubechies wavelet is used for ECG signal processing.

After the selection of a particular wavelet, ECG signal is decomposed using discrete wavelet transform. As ECG signal is decomposed in to six detailed coefficients then Entropy criteria is used in this paper to select one detailed coefficient. The detailed coefficient which shows highest energy with lowest entropy is chosen for feature extraction and classification.

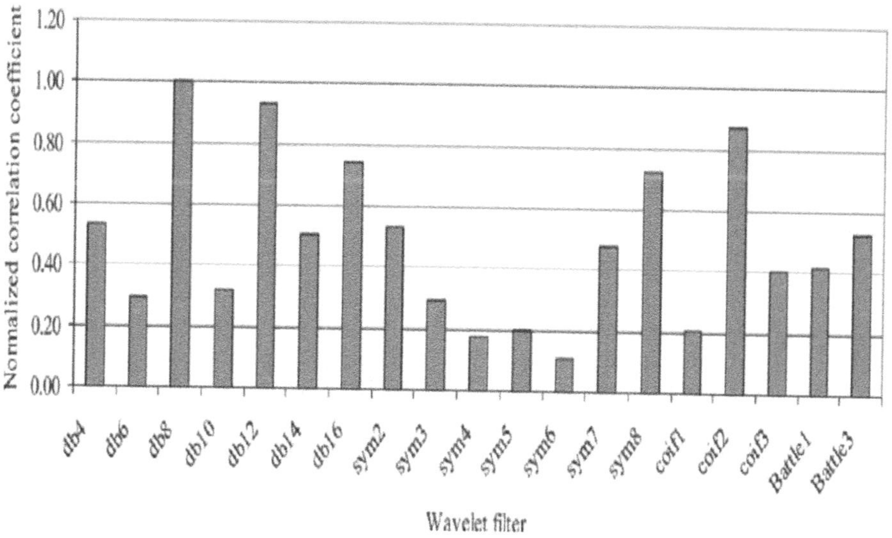

Fig. 2. Correlation between different wavelet and ECG signal

2.3 Feature Extraction

Human can perform the process of analysing and extracting Morphology and statistical features of ECG signals. Wave peaks (amplitude) as well as time durations(intervals) consist of clinically useful information present in ECG signal [6].

Morphological features were found as the most important for arrhythmia classification. Morphological features are calculated by signal processing methods while statistical features are calculated mathematically. In this paper, features which have been extracted are R wave amplitude, RR interval, standard deviation, variance, etc. These features are calculated by given formulae.

$$HR = 60/RR\,\text{Interval} \tag{1}$$

$$RMS = \sqrt{\frac{1}{N}\left(\sum x^2(n)\right)} \tag{2}$$

$$Varience = \frac{1}{N}\left(\sum_{n=1}^{N} [x(n) - x]^2\right) \tag{3}$$

$$SD = \sqrt{\frac{1}{N}\left(\sum_{n=1}^{N} [x(n) - \overline{x}]^2\right)} \tag{4}$$

2.4 Artificial Neural Network (ANN)

ANN used in this paper has 9 neurons in input layer, 1 hidden layer with 15 neurons and 3 neurons in output layer. Learning rate of 0.2, momentum factor of 0.5, error goal of

10^{-5}, 100 epochs and batch size of 1 were used during the training. Mean Square Error (MSE) has been used as the loss function of the ANN.

Once the features are extracted, normalization is done in such a way that each data should fall within ranging from 0 to 1[7]. As the values of features may vary, it is important to normalize all the features to same level and to obtain optimal results. Formula for normalization is given by Eq. (5),

$$NormalizedValue = 0.1 + \left[\frac{\{(ActualValue - MinimumValue)X0.8\}}{\{MaximumValue - MinimumValue\}} \right] \quad (5)$$

Back propagation algorithm is used for training the ANN. MSE is reduced between the real output (actual output) and desired output by adjustment of weights using gradient descent search by back propagation. Flow chart of back propagation algorithm is shown in Fig. 3.

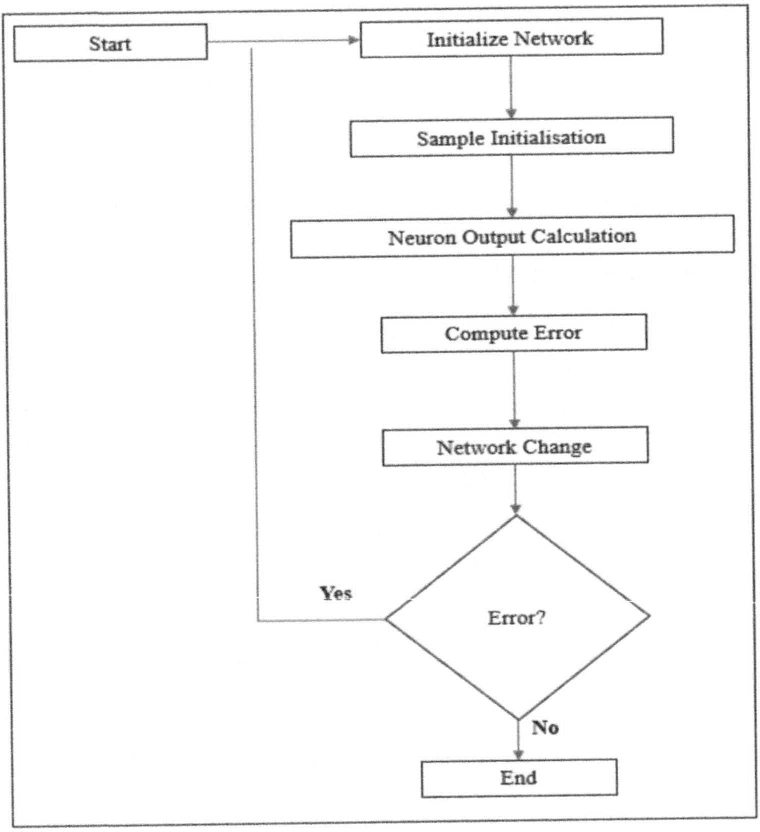

Fig. 3. Flow Chart of EBP algorithm

The basic steps used in back propagation algorithm are described below:

1. Find the different learning parameters based on predefined network structure, involving input, output and hidden layer, error measurements and learning rate[8].
2. Provide input and target vector.
3. Start learning and follow given steps:

Amount proportional to multiplication of error signal on the unit k and output of unit j adjust the weight of connection given below:

$$\Delta_p w_{jk} = \gamma \delta_k^p y_j^p \tag{6}$$

Error signal for output unit is given by equation:

$$\delta_o^p = (d_o^p - y_o^p) F'(s_o^p) \tag{7}$$

If sigmoid is the activation function then

$$y^p = F(s^p) = \frac{1}{1 + e^{-sp}} \tag{8}$$

Find the derivation of activation function

$$F'(s^p) = y^p(1 - y^p) \tag{9}$$

In such way that error signal is given for output unit

$$\delta_o^p = (d_o^p - y_o^p) y_o^p (1 - y_o^p) \tag{10}$$

a. Error signal for hidden layer is calculated in term of error signals of directly connected units and their weights. For sigmoid function:

$$\delta_h^p = {}'(s_h^p) \sum_{o=1}^{N_o} \delta_o^p w_{ho} = y_h^p (1 - y_h^p) \sum_{o=1}^{N_o} \delta_o^p w_{ho} \tag{11}$$

b. If we make changes in weight dependent of past weights with the help of adding momentum then oscillations can be avoided.

$$\Delta w_{jk}(t + 1) = \gamma \delta_k^p y_j^p + \alpha \Delta w_{jk}(t) \tag{12}$$

Every time new sample is entered till error requirement is not achieved. Re-random order is needed for input order during training [9]. When network response is obtained then error of the last output layer is found only, this error is recalculated according to instructions given on previous layers. The base of Neural Networks is back propagation algorithm[10]. Computation of error is done again and again for fast and accurate results. Error function is minimized in this algorithm.

3 Results and Discussion

3.1 Selection of Detailed Coefficient

In this paper, initially ECG signal is decomposed in to different detailed coefficients using DWT.A particular detailed coefficient is chosen on the basis of entropy criteria. Figures 4, 5 and 6 shows the entropy of different detailed coefficients for Normal, Supra ventricular (SV) and Atrial fibrillation (AF) respectively. It also indicates that detailed coefficient of level 3 for normal, level 1 for supra ventricular and level 3 for atrial fibrillation should be selected to extract features as it represents minimum entropy.

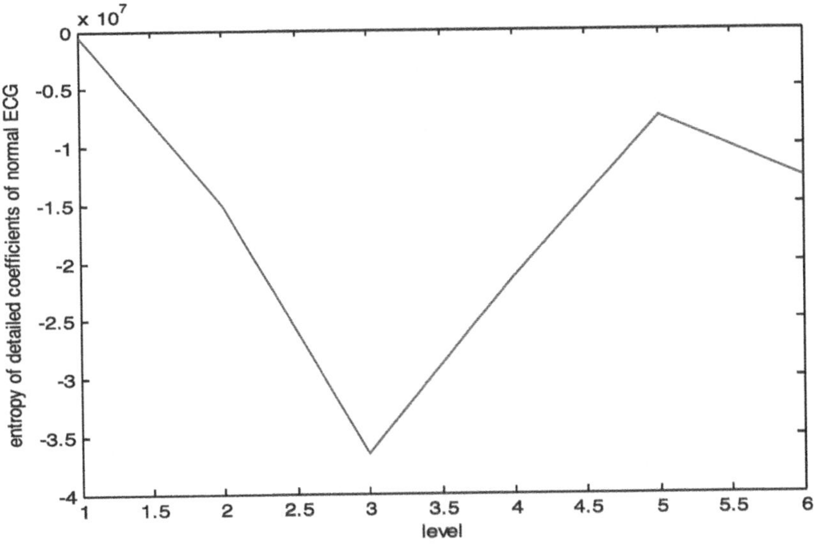

Fig. 4. Entropy Plot of Detailed Coefficient of Normal ECG

3.2 Feature Extraction & Normalization

There are two types of features (Statistical and morphological features) have been extracted for all three types of signals. After that they are normalized to obtain the optimum results. Table 2 shows the different normalized statistical features. These features provide information regarding cardiac health condition[11]. All these parameters are useful for discrimination between normal and arrhythmia patients. Different statistical features are calculated from normal, atrial fibrillation and supra ventricular ECG signals. Different statistical features calculated in this paper are maximum peak value, minimum peak value, mean, standard deviation, variance,entropy and mode.

Since most of the diagnostic information lies around the R peak of the ECG signal, hence a portion of signal before it and a portion of signal after it are selected for processing[12]. So After the extraction of statistical parameters, different morphological

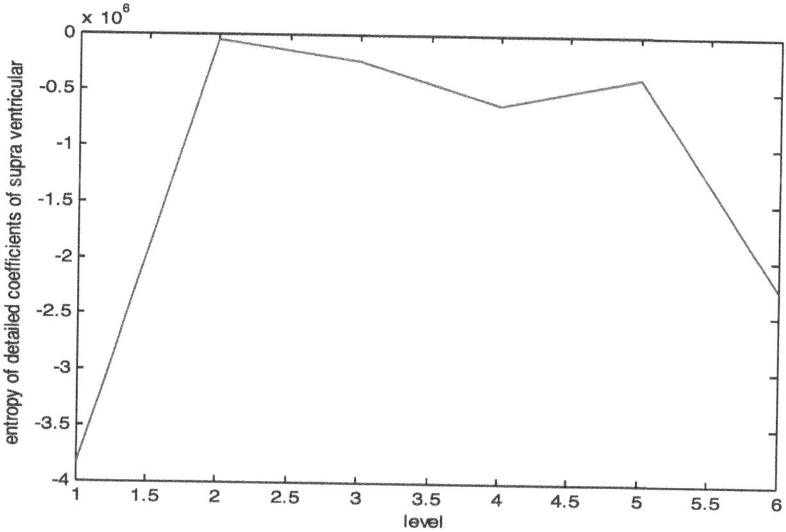

Fig. 5. Entropy Plot of Detailed Coefficient of Supra Ventricular

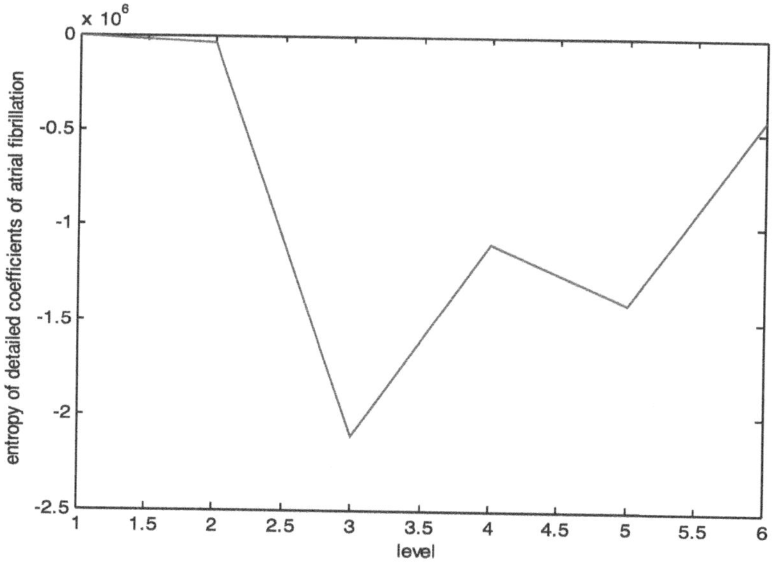

Fig. 6. Entropy Plot of Detailed Coefficient of Atrial Fibrillation

parameters like R wave amplitude, RR Interval are extracted.[13] Now the morphological features are normalized which are shown in Table 3.

R Peaks have been detected in normal ECG, super ventricular ECG and Atrial Fibrillation ECG and they have been marked by star using MATLAB (Figs. 7, 8 and 9).

Table 2. Normalized Statistical Features

Selected-detailed coefficients for Different ECG Signal	Mean (mv)	Mode	Entropy
Detailed coefficient of third level for normal ECG signal 1	0.1139	0.1012	0.2665
Detailed coefficient of third level for normal ECG signal 2	0.1134	0.1095	0.8686
Detailed coefficient of third level for normal ECG signal 3	0.1135	0.1000	0.2661
Detailed coefficient of third level for normal ECG signal 4	0.1135	0.1072	0.7745
Detailed coefficient of third level for normal ECG signal 5	0.1133	0.1085	0.8696
Detailed coefficient of third level for normal ECG signal 6	0.1134	0.1088	0.8813
Detailed coefficient of first level for SV ECG signal 1	0.1135	0.1084	0.1000
Detailed coefficient of first level for SV ECG signal 2	0.1136	0.1119	0.7965
Detailed coefficient of first level for SV ECG signal 3	0.1134	0.1118	0.5470
Detailed coefficient of first level for SV ECG signal 4	0.1134	0.1110	0.9000
Detailed coefficient of first level for SV ECG signal 5	0.1134	0.1120	0.8392
Detailed coefficient of first level for SV ECG signal 6	0.1135	0.1125	.8331
Detailed coefficient of third level for AF ECG signal 1	0.1134	0.1095	0.8632
Detailed coefficient of third level for AF ECG signal 2	0.1134	0.1085	0.8629
Detailed coefficient of third level for AF ECG signal 3	0.1134	0.1053	0.7669
Detailed coefficient of third level for AF ECG signal 4	0.1134	0.1103	0.8809
Detailed coefficient of third level for AF ECG signal 5	0.1134	0.1119	0.8959
Detailed coefficient of third level for AF ECG signal 6	0.1134	0.1087	0.8471

(*continued*)

Table 2. (*continued*)

Selected-detailed coefficients for Different ECG Signal	Mean (mv)	Mode		Entropy
Selected-detailed coefficients for different ECG signals	**Max. (mv)**	**Min. (mv)**	**SD (mv)**	**Var. (mv)**
Detailed coefficient of third level for normal ECG signal 1	0.123	0.101	0.116	0.754
Detailed coefficient of third level for normal ECG signal 2	0.117	0.109	0.114	0.163
Detailed coefficient of third level for normal ECG signal 3	0.125	0.100	0.117	0.900
Detailed coefficient of third level for normal ECG signal 4	0.119	0.107	0.115	0.282
Detailed coefficient of third level for normal ECG signal 5	0.116	0.108	0.114	0.195
Detailed coefficient of third level for normal ECG signal 6	0.116	0.108	0.114	0.165
Detailed coefficient of first level for SV ECG signal 1	0.117	0.108	0.114	0.152
Detailed coefficient of first level for SV ECG signal 2	0.117	0.111	0.114	0.136
Detailed coefficient of first level for SV ECG signal 3	0.115	0.111	0.113	0.119
Detailed coefficient of first level for SV ECG signal 4	0.116	0.111	0.113	0.125
Detailed coefficient of first level for SV ECG signal 5	0.115	0.112	0.113	0.118
Detailed coefficient of first level for SV ECG signal 6	0.118	0.112	0.114	0.139
Detailed coefficient of third level for AF ECG signal 1	0.118	0.109	0.114	0.163
Detailed coefficient of third level for AF ECG signal 2	0.119	0.108	0.114	0.183
Detailed coefficient of third level for AF ECG signal 3	0.120	0.105	0.115	0.301
Detailed coefficient of third level for AF ECG signal 4	0.116	0.110	0.114	0.152

(*continued*)

Table 2. (*continued*)

Selected-detailed coefficients for Different ECG Signal	Mean (mv)	Mode		Entropy
Detailed coefficient of third level for AF ECG signal 5	0.115	0.111	0.113	0.120
Detailed coefficient of third level for AF ECG signal 6	0.118	0.108	0.114	0.205

Table 3. Normalized Morphological Features

Selected detailed coefficient for different ECG signals	R Wave Amplitude (mv)	RR Interval (seconds)
Detailed coefficient of third level for normal ECG signal 1	0.8679	0.1095
Detailed coefficient of third level for normal ECG signal 2	0.4204	0.1259
Detailed coefficient of third level for normal ECG signal 3	0.900	0.1101
Detailed coefficient of third level for normal ECG signal 4	0.5503	0.1095
Detailed coefficient of third level for normal ECG signal 5	0.4641	0.1058
Detailed coefficient of third level for normal ECG signal 6	0.4600	0.1080
Detailed coefficient of first level for SV ECG signal 1	0.5145	0.1000
Detailed coefficient of first level for SV ECG signal 2	0.4537	0.1012
Detailed coefficient of first level for SV ECG signal 3	0.5551	0.1055
Detailed coefficient of first level for SV ECG signal 4	0.5076	0.1061
Detailed coefficient of first level for SV ECG signal 5	0.4378	0.1070
Detailed coefficient of first level for SV ECG signal 6	0.7434	0.1043
Detailed coefficient of third level for AF ECG signal 1	0.5044	0.1022

(*continued*)

Table 3. (*continued*)

Selected detailed coefficient for different ECG signals	R Wave Amplitude (mv)	RR Interval (seconds)
Detailed coefficient of third level for AF ECG signal 2	0.5076	0.1026
Detailed coefficient of third level for AF ECG signal 3	0.6173	0.1033
Detailed coefficient of third level for AF ECG signal 4	0.6066	0.1003
Detailed coefficient of third level for AF ECG signal 5	0.4759	0.1041
Detailed coefficient of third level for AF ECG signal 6	0.5139	0.1019

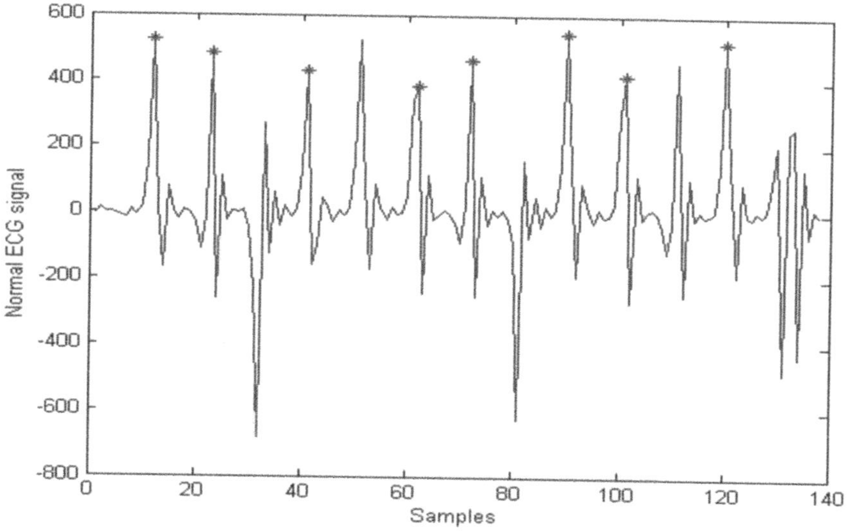

Fig. 7. Detection of R peak in Normal E

3.3 Back Propagation (BP) Algorithm

Multi layered feed forward network with BP algorithm is used during training and testing phase. Mean square error (MSE) is employed as error function [14]. But in this paper, Network is trained with momentum factor 0.5, learning rate 0.2, number of iterations 1000 and error goal 10^{-5}. As total selected decomposed coefficients are 18 and 9 features of each coefficient is determined so 9×9 matrix (9 features and 9 signals) are given as input of neural network for training purpose and another 9×9 matrix are

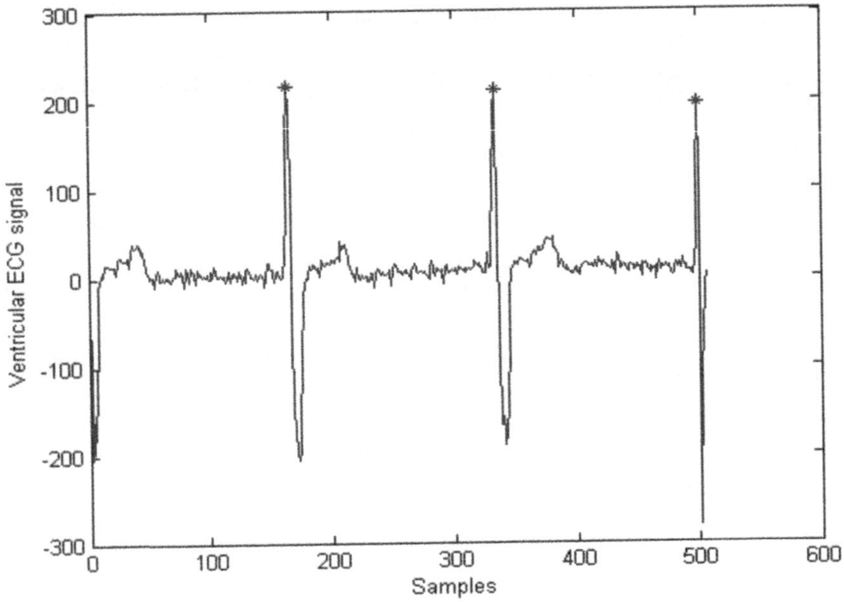

Fig. 8. Detection of R peak in Super Ventricular ECG

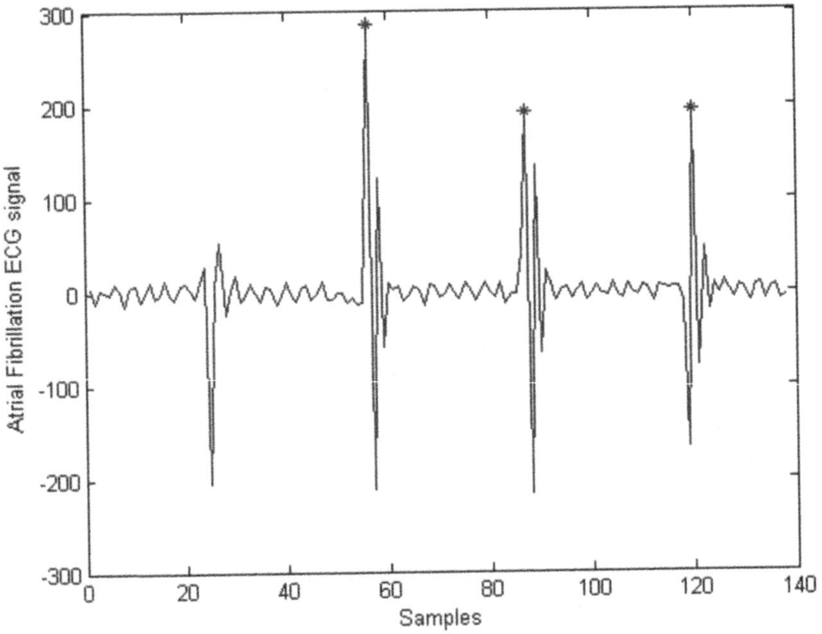

Fig. 9. Detection of R peak in Atrial Fibrillation ECG

used for testing purpose. Variation in MSE provides training performance with different iterations. Figure 10 shows the performance graph:

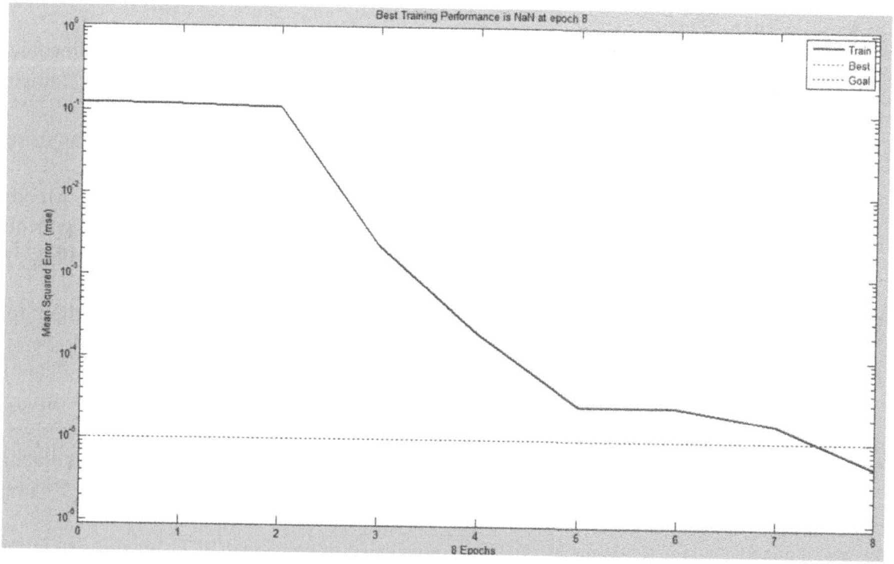

Fig. 10. Performance Analysis

The accuracy during the testing of the network is found to be 97.5% (only 2 out of 81 returned the negative results). The error performance goal is obtained at the eight epochs as shown in above figure. Accuracy can be increased by varying the network parameters (iterations, momentum factor, learning rate, number of inputs to the network and error goal).

4 Conclusion

The methodology used in this paper has been applied on 18 patients. The results of this work produced an accuracy of 97.5%.The result obtained in the paper shows that data base can be classified using ANN. Architecture for feature extraction of 18 ECG signals and their classification have been developed successfully.

This paper presents the use of wavelet transform and Multi-Layer neural networks for classification of normal and disease subjects. A classifier expert system for ECG signals can also be developed by using Fuzzy logic and Genetic algorithm. Features extracted from the program can be used for training purpose, concludes the dissertation indication for its utility in the field of Biomedical and future scope for further work. One step can also be added that features will be extracted by decomposition of ECG signal by WT, the feature selection will be done by Genetic Algorithm then more reduced features will be given to input of ANN for classification. Further, research is required in order to enhance in the field of clinical applications of signal processing methodology.

References

1. Zhu H, Pan Y, Wu F, Huan R (2019) Optimized electrode locations for wearable single-lead ECG monitoring devices: a case study using WFEES modules based on the LANS method. Sensors, **19**(20) (2019)
2. Hamidi AA, Robertson B, Ilow J (2023) A new approach for ECG artifact detection using fine-KNN classification and wavelet scattering features in vital health applications. Proc Comput Sci, **224**:60–67 (2023)
3. Singh R, Mehta R, Rajpal N (2018) Efficient wavelet families for ECG classification using neural classifiers. Proc Comput Sci, 11–21 (2018)
4. Goldberger A, Amaral L, Glass L, Hausdorff J, Ivanov PC, Mark R, Mietus JE, Moody GB, Peng CK, Stanley HE (2000) PhysioBank, PhysioToolkit, and PhysioNet: components of a new research resource for complex physiologic signals. Circulation [Online]. **101**(23), e215–e220
5. Das M, Sahana BC (2022) Optimized orthogonal wavelet-based filtering method for electrocardiogram signal denoising. J Inst Eng (India): Series B, 1–14
6. Montenegro L et al (2022) Human-assisted vs. deep learning feature extraction: an evaluation of ECG features extraction methods for arrhythmia classification using machine learning. Appl Sci.
7. Rahul J, Sharma LD (2022) Artificial intelligence-based approach for atrial fibrillation detection using normalised and short-duration time-frequency ECG. Biomed Sig Process Control
8. Hammad M, Zhang S, Wang K (2019) A novel two-dimensional ECG feature extraction and classification algorithm based on convolution neural network for human authentication. Future Gener Comput Syst, 180–196
9. Pandey, Kumar S, Janghel RR (2019) ECG arrhythmia classification using artificial neural networks. Proceedings of 2nd International Conference on Communication, Computing and Networking: ICCCN 2018, NITTTR Chandigarh, India. Springer Singapore
10. Bishop, A.J., Nehme, Z., Nanayakkara, S., Anderson, D., Stub, D., Meadley, B.N.: Artificial neural networks for ECG interpretation in acute coronary syndrome: a scoping review. Am. J. Emerg. Med. **83**, 1–8 (2024)
11. Asgharzadeh-Bonab A, Amirani MC, Mehri A (2020) Spectral entropy and deep convolutional neural network for ECG beat classification. Biocybernetics Biomed. Eng., **40**(2):691–700
12. Mandal S, Mondal P, Roy AH (2021) Detection of ventricular arrhythmia by using heart rate variability signal and ECG beat image. Biomed Signal Process Control, **68**:102692
13. Fan, W., Liu, K., Liu, H., Ge, Y., Xiong, H., Fu, Y.: Interactive reinforcement learning for feature selection with decision tree in the loop. IEEE Trans. Knowl. Data Eng. **35**, 1624–1636 (2021)
14. Alotaiby, T.N., Alrshoud, S.R., Alshebeili, S.A., Aljafar, L.M.: ECG-based subject identification using statistical features and random forest. J Sens **2019**, 1–13 (2019)

Analyzing the Hybrid Electric Vehicle's Performance for Automatic Acceleration Control

M. L. N. Charyulu, M. Sailaxmi, and Vivek Singh Kushwah(✉)

Department of Electronics and Communication Engineering, Chaitanya Bharathi Institute of Technology (Autonomous Institute Affiliated to Osmania University), Gandipet, Hyderabad, Telangana 500075, India

{mlncharyulu_ece,viveksinghk_ece}@cbit.ac.in

Abstract. Hybrid electric vehicles primarily revolve around improving fuel efficiency, reducing emissions, addressing environmental concerns, and offering a more accessible transition toward electric mobility while leveraging existing automotive technology. HEVs are designed to combine the benefits of traditional internal combustion engines (ICE) with those of electric propulsion systems, in heightened total vehicle performance, decreased emissions, and increased fuel efficiency. The development of a driver, electric powertrain, and vehicle model for a hybrid electric vehicle. Examine the conduct of the Plot the real velocity power distribution curves for the car throughout a US06 drive cycle. Examine the fuel consumption and operation of the battery, ultracapacitor, and fuel cell as well. By converting to power gearbox via the motor and turning off the engine, the HEV resolves the issue. In this manner, no fuel will be used when the vehicle is idle, producing no exhaust emissions. A significant benefit of hybrid electric vehicles (HEVs) is their ability to operate entirely on electricity when the fuel tank runs empty while the engine is running.

Keywords: HEV · VAE · ICE · SOC · OCV

1 Introduction

An electric propulsion system replaces an internal combustion engine (ICE) in an electric vehicle, or EV. The vehicle's entire power is derived from an electric motor, which also acts as its primary drive source. The superior power conversion efficiency of the electric motor propulsion system is the primary benefit. Numerous reports of research and development activity relating to the business have surfaced recently. Many governments have given consumers incentives as a result of the increasing use of electric vehicles, such as lower taxes or tax refunds, free parking, and free or affordable charging stations. A hybrid electric vehicle (HEV) is a wise option, on the other hand. In the last few years, it has received a lot of attention. Nearly every significant automaker in the globe has at least one fully electric or hybrid vehicle available. Five major worldwide trends are contributing to the explosive expansion of hybrid electric vehicles, or HEVs: Depletion of fossil

A. Bagwari et al. (Eds.): ICEWTS 2024, CCIS 2399, pp. 23–48, 2025.
https://doi.org/10.1007/978-3-031-87886-2_3

fuels, A rise in public consciousness, Technological developments, the advancement of electronic control systems and electric motors and Developments in technologies that support EVs. Transportation contributes significantly to CO_2 emissions; in 2020, it will be responsible for 22% of world emissions. EVs, on the other hand, are less susceptible to rising oil prices and do not generate CO2. In the next 50 years, there will bean increase in the global population from 7 billion to 11 billion people, and there will be 2.5 billion more cars on the road than there were 700 million. As a result of a growth in ICE consumption, it is predicted that fossil fuels will run out by 2050. If gasoline and diesel engines are used, pollution will increase and the sky will become dark and dull. Forcing people to work for efficient transportation in the 21st century is one of the solutions to a warmer, cleaner future. HEV is the best hybrid car. Power drive by electric motors and I.C. combines with. Engine. The generator is powered by the vehicle's battery. In HEV, the IC engine is combined with the electric motor, allowing the engine to be used more efficiently. City driving requires frequent starting and stopping. During idling, the engine uses more fuel than it operates efficiently. This leads to higher fuel consumption, lower efficiency and unnecessary emissions. Hybrid cars solve this problem by switching power transfer from the engine and turning off the engine. In this way, fuel consumption and exhaust gas will not occur during operation. Another important advantage of HEVs is that the vehicle can increase its electric power to its maximum speed when the fuel tank is empty when the engine is started. The Specific Objectives of the Research paper is to observe the various plots of the Vehicle for the drive cycle such as Acceleration/Brake, Torque required, Power required, Power distribution among three energy sources, Fuel cell voltage, current, power and fuel consumption and Energy Management distribution algorithm for positive power distribution. The Energy management system implemented here will follow rule- based algorithm for the implementation of series parallel hybrid systems. The operating range of the sources are Battery and fuel.

1.1 Related Work

In this section, we review the existing literature on novel technologies implemented for Hybris electric vehicles. The idea and construction of an experimental hybrid electric car that combines solar energy from photovoltaic cells with human assistance are presented in this study. The vehicle is equipped with batteries to store the energy supplied by the two engines. The purpose of this application is to use the photovoltaic module as a large energy source to operate the vehicle with the least amount of energy usage possible. The South American solar automobile helped in the development of the solar car that traveled almost 1000 km in the Atacama desert in Chile, South America. Key points, power issues, and experimental results are presented [1]. This paper deals with most cars on the road now use fossil fuels in the form of petrol or diesel. Fewer people drive electric or electric hybrid cars, which are more environmentally friendly because they have less energy. Many of these issues have been resolved by supercapacitors, and the automotive industry now prioritizes research in this field as it relates to motor vehicles. Supercapacitors are described in this article along with their advantages. Along with a comparison between supercapacitors and ordinary batteries, it also explains how supercapacitors are being employed in electric and hybrid cars and explains their advantages over current batteries [2]. In [3] conducted a comprehensive review of recent developments and previous research in collaborative storage over the past two decades. The main objective of this study is to present a research proposal in which each category and subcategory is clearly and independently defined. The use of energy storage technology is one of the most interesting and beneficial for further development in the field of other technologies. In addition to careful storage-cooling of energy, the main goal is to improve the quality of storage facilities to achieve technological economic and practical studies. This paper examines hybrid energy storage consisting of batteries and supercapacitors. In[4]Presented Plug-in Hybrid Electric Vehicles (PHEVs) have numerous opportunities with charging devices and charge storage systems. With proper design and construction, electric vehicles (EVs) not only eliminate pollution, but also make the body more efficient than traditional cars. However, the investigation is ongoing and there are several unanswered questions. After then, the adaptive equivalent consumption minimization strategy (A-ECMS)-based lower level controller controls the energy management of the HEVs based on the optimal velocity profile. Meanwhile, for the purpose of computing the velocity profile for the upcoming time window, the higher-level controller fixes the fuel consumption model and updates the efficiencies with the average values. For the current time window, the average propulsion and recuperation efficiencies are calculated and fed back by the lower level controller. Results from the simulation confirm the control efficacy of the proposed control approach [5]. The absence of standardization in parts and solutions is impeding the introduction of electric vehicles into the mainstream market. The environmental and societal implications of this technology make standardization particularly crucial. However, because of the interconnected technologies and parties involved, it is also especially difficult [6]. In order to improve speed control and energy economy, this survey study examines the driving and braking control logic algorithms in electric cars (EVs). It highlights the possible effects of combining EVs with renewable energy technologies as it examines their applications with Smart Grids and Sustainable Energy systems. Additionally, it looks at new developments in adaptive control strategies to maximize the energy efficiency of electric cars (EVs). It concentrates on important parts such as energy management systems, regenerative braking, and the powertrain. With the use of these techniques, precise control is possible, considering variables like traffic volume, battery life, and road conditions. The survey study also looks into how EVs may be integrated with Smart Grid infrastructure, emphasizing the benefits of using renewable energy sources for EV charging and vehicleto-grid communication. Additionally, this research provides a thorough analysis [7]. This study looks at the WTW efficiency of EVs, compressed natural gas vehicles (CNGV), and conventional internal combustion engine cars (ICEVs) (gasoline, diesel). The findings indicate that WTW is significantly impacted by power plant efficiency, Effectiveness [8],EVs cause a change in the non-exhaust PM composition from brake to tyre wear. Compared to lesser weight EVs, heavier weight EVs may release higher non-exhaust PM2.5. Effectiveness[9].The paper specifically offers an assessment of how the development of future EVs—such as shared mobility, autonomous driving, and connected cars—will impact the integration of EVs into the grid and how the power grid's progress towards the future energy Internet will be impacted by and benefit from EV development. Lastly, an assessment and summary of the recommendations and issues for the advancement of the infrastructure related to EV charging and grid integration are provided. [10].For long-lasting electric cars, accurate lithium-ion battery lifespan calculation and health assessment are essential. An early detection of poor performance makes timely maintenance of battery systems easier Accidents and malfunctions are avoided, and operational costs are reduced. Recently, there has been increased interest in data-driven battery health estimates due to developments in "Big Data" analytics and associated statistical and computational techniques [11]. The fundamental mapping between capacity, storage temperature, and state-of-charge is captured using the Gaussian process regression (GPR) approach. High relevant input characteristics may be efficiently retrieved to increase prediction accuracy and resilience by altering the isotropic kernel function with an automated relevance determination (ARD) structure. Rather than utilising the data from a single scenario, the model is trained, validated, and compared using experimental battery calendar ageing data from nine storage situations. This approach is more relevant and useful. Results show that the suggested GPR model with ARD Matern32 (M32) kernel performs better than its alternatives and can produce accurate predictions for any storage scenario.[12], This article [13] offers a thorough analysis of the fault characteristics, fault processes, and fault diagnosis of a variety of LIBS defects, such as actuator, sensor, and internal battery issues. The objective is to encourage creative thinking for LIBS fault diagnostics and to advance a thorough grasp of the newest technology.

There isn't a single diagnostic technique that works for every battery system malfunction; all of them require more investigation and refinement. Battery malfunctions exhibit several modes, intricate systems, and interdependent interactions. The predicted battery condition and model parameters, including capacity, internal resistance, SOC, and temperature, usually fluctuate abnormally as a result of these defects. Thus, the most widely used techniques for diagnosing battery faults are now model-based state estimates and parameter estimation.

1.2 Motivation and Contribution

Inspired by the previous works and promising characteristics of Hybrid Electric Vehicles and attention mechanism in the Real field, In this paper, we propose a unique Performance Investigation of Hybrid Electric Vehicle using Matlab Simulink. It may reduce your carbon footprint by driving an electric automobile as they don't produce any exhaust emissions. You may further reduce the environmental impact of automobile charging by choosing renewable energy sources for your home's power. With the extra power the electric motor offers, a smaller engine would be feasible. Furthermore, while the engine is stopped, the battery can power auxiliary loads and reduce engine idle. When combined, these features increase fuel economy without sacrificing functionality. Learn further information on hybrid electric automobiles. The Methodology of work is to develop an analysis project that can model the power flow of a HEV drivetrain and calculate the characteristics of a specific vehicle when performing various tasks. Knowing the required energy, the model allows us to determine the location of the allowed states at each time step, which becomes: It is a combination of engine power and corresponding engine power, considering regenerative braking and the limitations imposed by the engine, motor and storage systems. MATLAB SIMULINK Software tools are used for implementing results. This paper is divided into the following sections: Section I describes about the introduction after that about the literature survey and motivation and contribution related to the work. Section II describes about the theory related to the work. Section III describes about the implementation of the work, later on the experimental setup and simulation results are given in section IV and at last section V contains conclusion of the research work. The main contributions of the papers are summarized as follows. As to optimizing Retrieve energy and retain it within the battery. Occasionally, turn off the engine. Utilise cutting-edge aerodynamics to lower drag. Put on tyres with reduced rolling resistance. Make use of light materials.

2 Concept of Vehicles

As we all know, an electric motor is what defines a "EV" as any type of vehicle. Electric vehicles fall under this wide category and include hybrid electric vehicles (HEVs), fuel cell electric vehicles (FCEVs) and battery electric vehicles (BEVs) (Fig. 1).

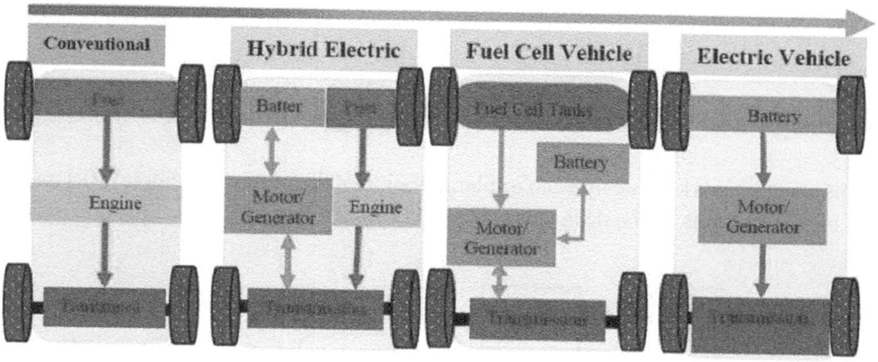

Fig. 1. Difference between Conventional and various EVs

2.1 Advantages and Disadvantages of HEV

The following are the benefits of a hybrid electric car over a traditional internal combustion engine:

(i) Lower energy losses: By automatically shutting down the engine when it is not in use, the hybrid system saves energy that would otherwise be squandered.
(ii) Energy recovery: During braking and deceleration, the vehicle's kinetic energy which would often be squandered as heat is converted into electrical energy that the electric motor uses later. (iii) Electric motor assistant: By assisting the engine with acceleration during transient operations, the electric motor enhances dynamic responsiveness and lowers exhaust pollutants. (iv) Extremely efficient motor operation: The motor's operating point (torque and speed) can be kept within the most

cost-effective range by utilizing an electric machine or machines as a motor or generator. (v) Driving just on electricity: The car may operate in electric-only mode at low speeds, which results in no fuel consumption or exhaust emissions. In contrast to internal combustion engines, electric motors have a great energy efficiency and can deliver torque instantly. These benefits enable the electric motor to support the internal combustion engine (ICE) during acceleration phases and to temporarily increase torque.

Hybrid electric vehicles' (HEVs') drawbacks include:

- Necessitates the incorporation of a fully automated gearbox.
- The torque of an electric machine is only amplified by half of the gear ratios, whether even or odd.
- Because of an electric machine or permanent connection, coasting is not feasible.
- Needs more room on the rear axle for installation.
- Needs two electrical devices with comparable power ratings.
- The torque and power of the electric machine are the only factors that affect the vehicle's dynamic performance.
- Energy-inefficient because motor power is converted twice (mechanically, electrically, and mechanically).

3 Design and Implementation of Proposed Vehicle

In this research work the design and implementation of the proposed vehicle is discussed. The block diagrams of the design are shown in Fig. 2.

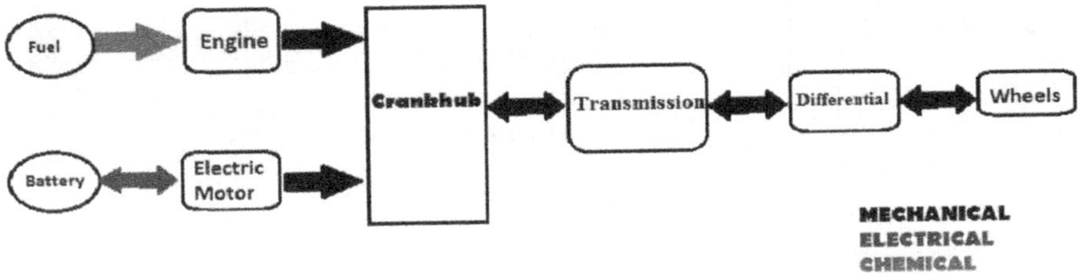

Fig. 2. Flow chart for Hybrid Architecture

A hybrid car employs two energy sources and two energy converters when looking at its operating principles. These regulations control how an electric hybrid car operates. The energy sources are divided into two categories: main and secondary. A hybrid electric vehicle's (HEV) battery serves as its secondary energy source, while the petrol tank is its primary source. There is a significant difference in the energy content of the primary and secondary energy sources. Though not the other way around, energy can go from the primary to the secondary energy source. Energy converters are used to transmit energy from the primary source to the secondary source. In a hybrid energy vehicle (HEV), the internal combustion engine serves as the primary energy converter, while the electric machine (motor/generator) serves as the secondary energy converter. Only the secondary energy converter is capable of recovering and storing a portion of the vehicle's kinetic energy while braking the backup energy supply. The wheel may receive simultaneous traction torque application from both energy converters. The complete Simulink model of the vehicles shown below.

There are 3 main subsystems: Driver, Electric power train and Vehicle. Each of these sub systems modeled in Simulink is explained in Fig. 3.

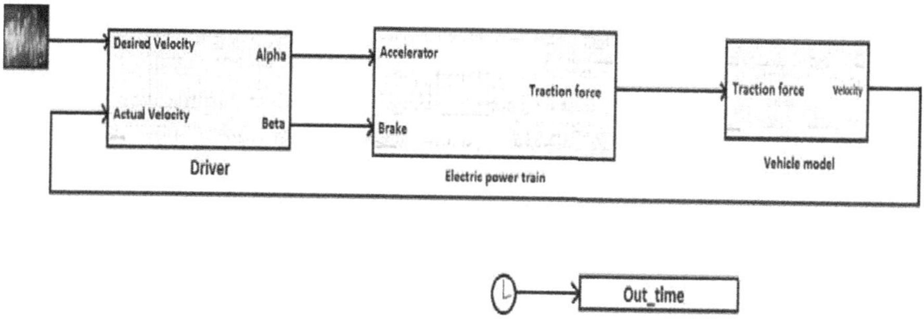

Fig. 3. Basic vehicle model

3.1 Driver Model

The basic function of driver block is to generate Acceleration.

(Alpha) and Brake (Beta)signals according to Desired velocity and Actual velocity commands as illustrated in Fig. 4.

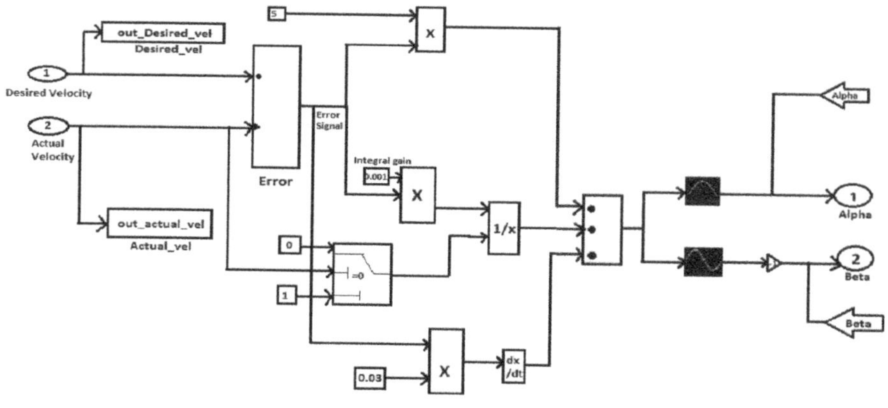

Fig. 4. Driver

The input to the driver model is desired velocity and actual velocity. The desired velocity is the driving cycle input provided through a repeating sequence of which velocity and time profiles are declared in workspace. In our model, the input is given by US06 drive cycle.

In order to solve the FTP-75 test cycle's deficiencies in representing aggressive, high speed and/or fast acceleration driving behaviour, rapid speed fluctuations, and driving behaviour after startup, the US06 Supplemental Federal Test Procedure (SFTP) was created. The vehicle's actual velocity is determined by using output as a feedback. The error signal which is the difference between desired and actual velocity is fed through a PID controller. If the integral block does not reset properly, it may cause integral wind-up problem. It occurs due to the limitation in actuators. The correction from the integral term of PID controller increases a lot which causes the system output to go beyond the setpoint. It is known as overshoot, and it increases with time. To avoid these problems, external reset is provided to the integral block. Whenever the velocity becomes zero, this integral block gets reset. In Table 2, It is implemented through a switch. It passes a zero to the integral block whenever velocity gets zero. The output of the PID controller is fed to the saturation block to get positive values. The Alpha can have any value from 0 to 1 and Beta from 0 to -1. The values of PID controller are optimized by observing the output velocity and input velocity profiles (Table 1).

Table 1. Comparison of different types of Electric Vehicle.

Types of EV's	Battery EV's	Hybrid EV's	Fuel Cell EV's
Propulsion	• Electric motor drives	• Electric motor drives • Internal combustion engines	• Electric motor drives
Energy system	• Battery • Ultra-capacitor	• Battery • Ultra-capacitor • ICE generating unit	• Fuel cells
Energy source & Infrastructure	• Electric grid charging facilities	• Gasoline stations • Electric grid charging facilities(optional)	• Hydrogen • Methanol or Gasoline • Ethanol
Characteristics	• Zero emission • Independence on crude oils • 100-200km short range • High initial cost • Commercially available	• Very low emission • Long driving range • Dependence on crude oils • Complex • Commercially available	• Zero emission or ultra -low emission • High energy efficiency • Independence on crude oils • Satisfied driving range • High cost now • Under development
Major issues	• Battery and Battery management • High performance Propulsion • Charging facilities	• Managing multiple energy sources • Dependent on driving cycle • Battery sizing and management	• Fuel cell cost • Fuel processor • Fueling system

Table 2. Observation.

S.No.	Condition	Acceleration	Brake
1	Desired velocity > Actual velocity	1(Vehicle is moving)	0
2	Desired velocity < Actual velocity	0	1(brake is applied)
3	Desired velocity = Actual velocity	0(idle)	0(idle)

3.2 Electric Power Train

The Electric power train contains the blocks such as Controller Power Calculation, Tractive Force, EMS (Energy Management System), UC model, Fuel cell, Battery model.

The input/output signal arrangements inside 'Electric power train' subsystem is as shown Fig. 5.

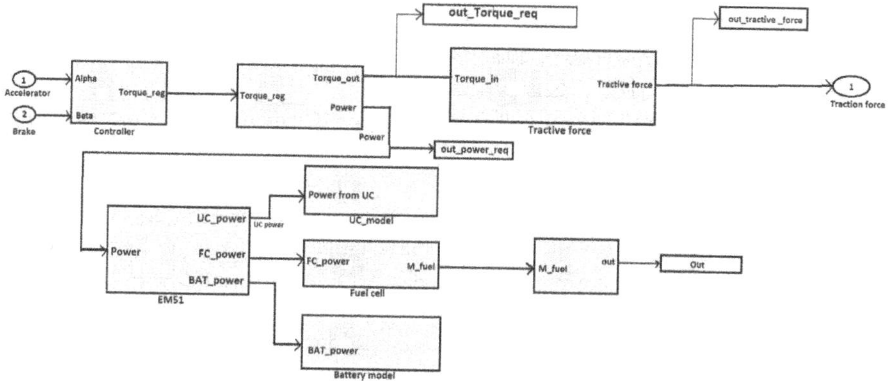

Fig. 5. Electric Power Train

Controller:

The controller block generates the required torque signal from Acceleration and Brake commands. Electric motor torque (positive) values are declared in workspace as variable 'EM_ new. Max T_ T' array and negative values are declared as variable 'EM new. Min T_T' array as shown in Fig. 6.

The Acceleration and brake outputs of 'Driver' subsystem is fed to 'Controller' block. The 'max' block will take the maximum value and 'min' block will take the minimum value. The SOC of ultra-capacitor is being checked for regenerative braking. If the SOC_UC > 0.8, then it won't be further charged while braking. If the SOC_UC < 0.8, then UC will get charged when braking occurs. While Acceleration positive torque request will be generated and while braking, negative torque will be generated. This torque request is fed to 'Power Calculation' block.

Power Calculation:

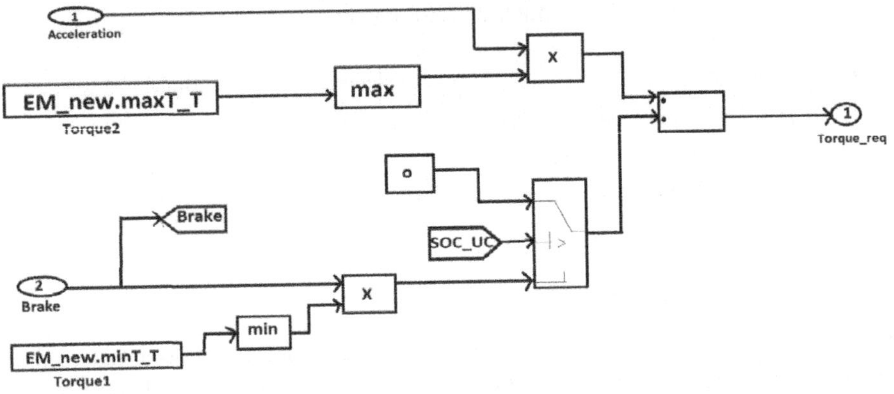

Fig. 6. Controller

This subsystem will calculate the required power for a given angular velocity and Torque. To calculate power, we need angular velocity, torque, and efficiency.

Power = (Torque*Angular velocity)/ Efficiency.

Torque

For a given acceleration/brake the required torque being calculated by Controller subsystem and fed to the 'Power calculation' block. Inside this there is a dynamic saturation block which will check whether the required torque is in the limits of torque that can be supplied by the Electric Motor. This is done by comparing the 'Torque req' with two curves of 'w' versus 'Torque'. Where 'w' is the Angular velocity in 'rad/s'. One curve gives the values of positive torque for particular 'w' and other one gives the values of negative torque. While acceleration, torque will be positive and while braking, negative.

This dynamic saturation block will give the torque request value which can be supplied by the electric motor. This is given as 'Torque out' signal.

3.2.1 EM Efficiency

This table will do some extrapolation and interpolation methods to give the value of Electric Motor efficiency for a particular toque and angular velocity. This can give efficiency values for torque and w values which are not listed in the table. A Saturation block is provided at the output of this block to ensure that the efficiency value is positive.

If the 'Torque req' is positive (Acceleration).

Then Power = (Torque * Angular velocity) /Efficiency If the 'Torque req' is negative (Braking) Then Power = (Torque*Angular velocity) *Efficiency. Since the motor will act in "Generator mode'.

3.2.2 Tractive Force

The Torque to Tractive force calculation can be done in 'Tractive force' subsystem which includes Gearbox model, Front brakes, Front wheels, Rear brakes and Rear wheels. The total tractive force generated will be the sum of forces in front wheels and rear wheels. We consider that electric motor is coupled to the front wheels and so the braking torque will be provided to front wheels alone. The rear wheels will only have mechanical braking. The 'Tractive force' subsystem is as shown below in Fig. 7.

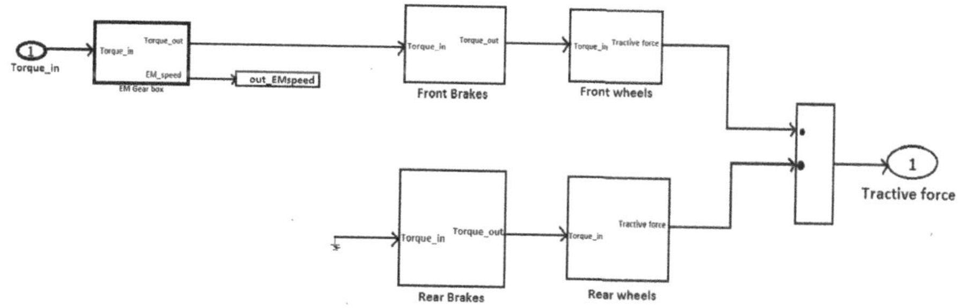

Fig. 7. Tractive Force

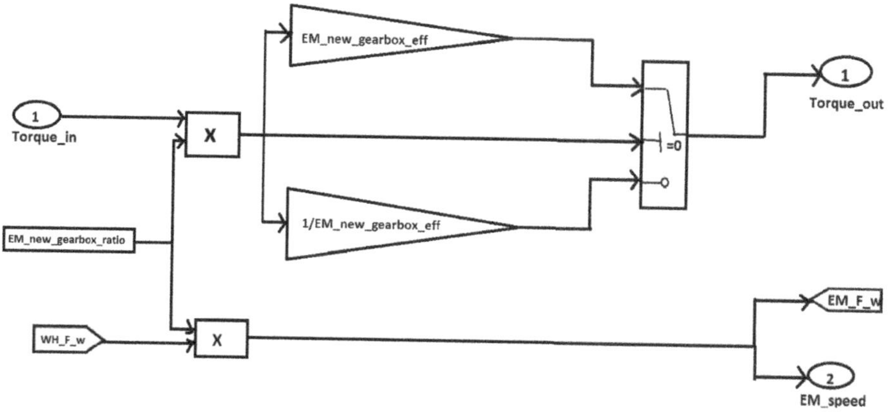

Fig. 8. Gear Box

The output of the gear box depends upon the torque input from electric motor and the efficiency of the gear box as shown in Fig. 8.

The output of the gear box,

Torque out = Torque * Gear box ratio * Efficiency [For + ve Torque] Torque out = Torque * Gear box ratio / Efficiency [For -ve Torque].

The electric motor speed, EM speed = Gear box ratio * wheel speed The wheel speed (angular velocity) is calculated in the front wheel model. Figure 9 illustrates the block diagram of Front Brakes Model.

The traction force in front wheels is provided by regenerative braking and mechanical braking.

The braking torque is given by,

Braking Torque = Brake signal * vehicle mass * ground friction * gravity * tire radius. The front and rear wheels model is as shown in Figs. 10 and 11 respectively.

The tractive force can be calculated as Tractive force = Torque / radius of tire.

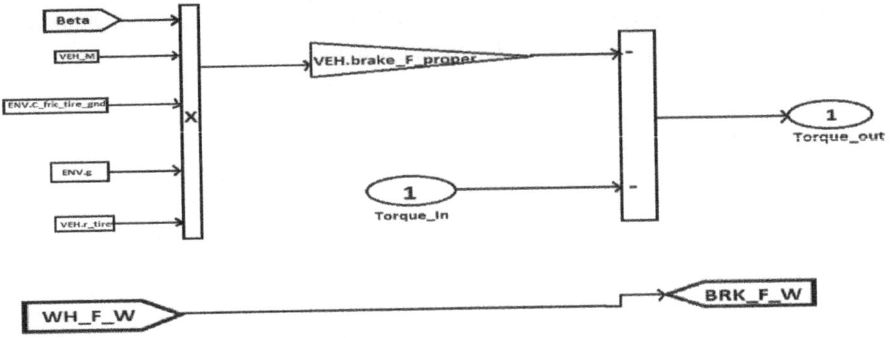

Fig. 9. Front Brakes

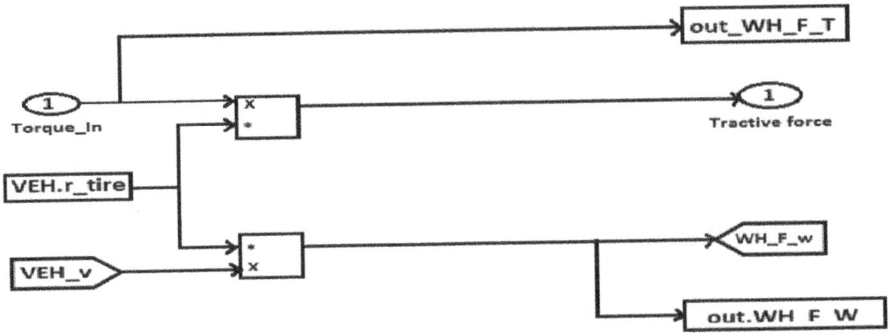

Fig. 10. Front wheels

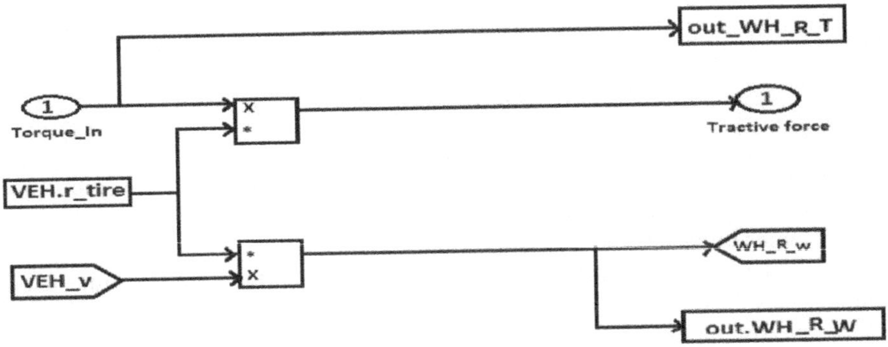

Fig. 11. Rear Wheels

The angular velocity of the vehicle calculated as Angular velocity = Vehicle velocity / radius of tire.

This is being fed to the gear box block to find the electric motor speed.

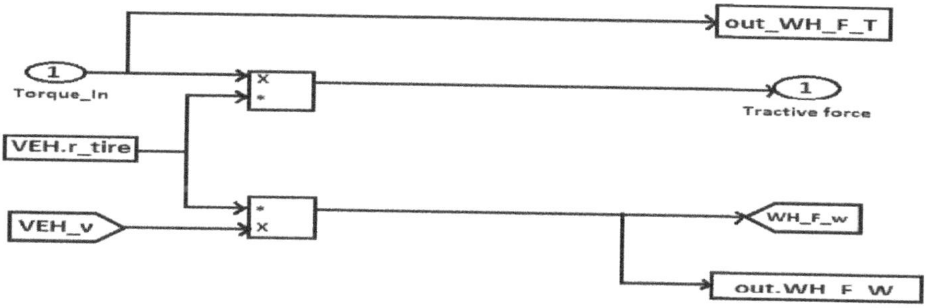

Fig. 12. Rear Brakes

Only mechanical braking is applied to rear wheels. So, the torque input is connected to the ground and the tractive force is calculated similarly as front brakes block in Fig. 12. Since the Torque input is zero, there won 'be any tractive force generated.

3.2.3 Energy Management System

The Energy Management system should allow energy sources to operate in their optimum performance range and at the same time it should be able to provide power even if no sources are available. Hybrid energy management systems have the advantage of providing uninterruptible power. In our case we are considering three energy sources for power distribution. The operating range of each of the sources and Backup power distribution is as follows as shown in Fig. 13. Inside 'Positive power subsystem, Power has been compared using interval test block and switches. The primary power distribution is as shown in the table above.

Inside the 'Positive power' subsystem, there is another subsystem called 'Backup power'. Inside that the battery SOC and fuel cell SOC have considered, and backup power options are provided for each condition.

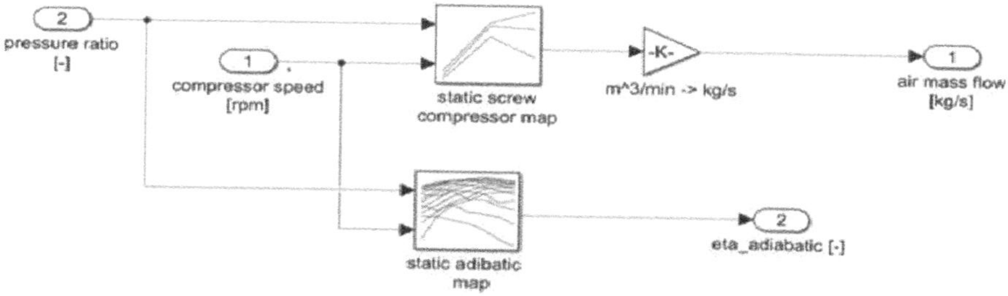

Fig. 13. Compressor model subsystem

The air flow rate from the compressor depends on the Pressure ratio, Compressor speed and Adiabatic efficiency. The adiabatic efficiency map gives the efficiency of the compressor for different compressor speed and pressure speed as shown in Fig. 14.

In Fig. 15, the compressor output power can be calculated as follows: P_comp = (W_comp * T_amb * Cp) * (((P2/P1) (1.4–1/1.4) -1) / eta_adiabatic).

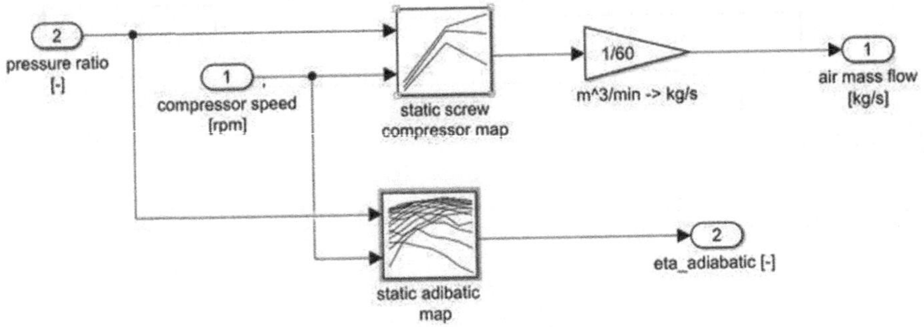

Fig. 14. Static screw compressor map

Where W_comp = air flow rate[kg/s] T_amb – Ambient temperature Cp - specific heat at constant pressure (J/gK) P2 – Raised pressure. P1 – Atmospheric pressure.

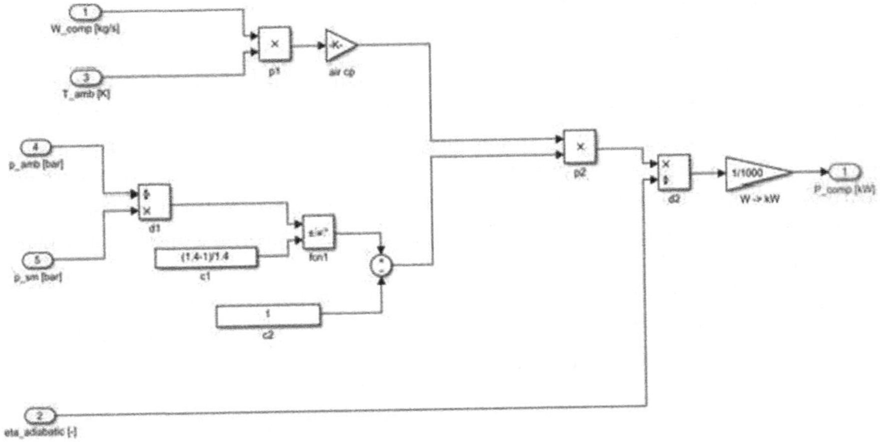

Fig. 15. Power

Battery Model

The Li-Ion battery model is as shown in Fig. 16. The input to the model is the power request from the EMS. There are two subsystems inside it. a. Current calculation.

b. SOC calculation.

The current calculation subsystem is as shown in Fig. 17. Battery voltage is determined from SOC. The open circuit voltage with SOC data for a single cell will be available from the datasheet. It must be multiplied with total number of cells in series to get the total voltage of the battery pack.

There are 3 plots used in this model to calculate the current. The open circuit voltage of the battery is determined from the SOC Vs OCV plot.

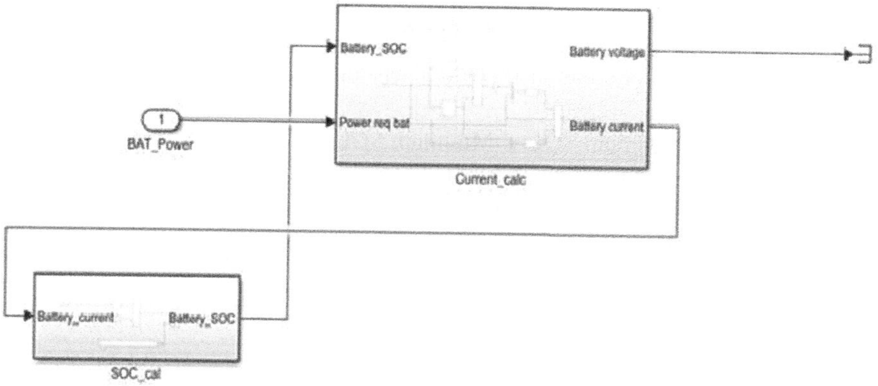

Fig. 16. Battery model

The internal resistance of the battery will be different while charging and discharging. This value can be obtained from SOC Vs Resistance for discharging and charging curves. Then current, I = sqrt(Power /R(chg/disch)).

The battery terminal voltage = Battery power / current.

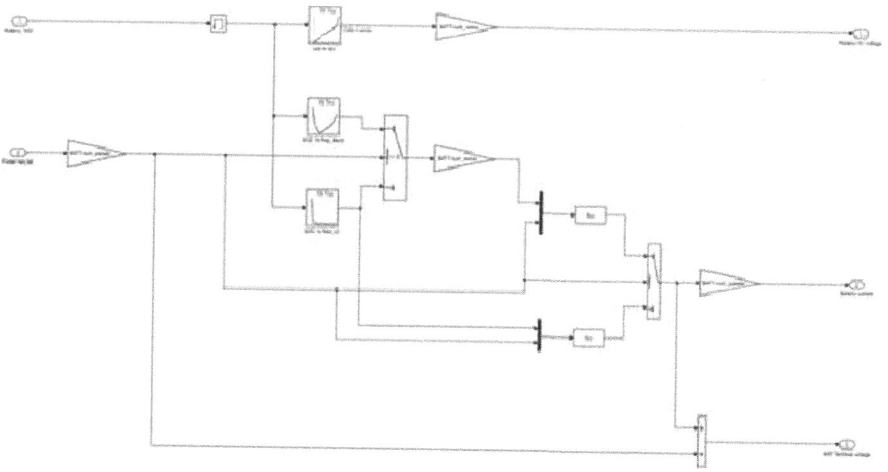

Fig. 17. Current calculation subsystem

The battery's estimated state of charge is displayed in Fig. 18. There are several ways to calculate a battery's state of charge (SOC). The coulomb counting approach is the one used in this instance. In order to determine SOC, the Coulomb counting method analyses a battery's discharging current and integrates it over time. The SOC (t) is computed using the Coulomb counting approach, which is based on the discharging current, I(t), and the previously determined SOC values, SOC(t – 1). This formula is used to determine SOC: SOC(t) = SOC(t-1) + (I(t) * Δt) / Qn.

Where Qn is the total Ah capacity of the battery. Here it is given as a constant value of 200Ah. Which is converted to mAh and fed to the block. There should be an initial SOC for the battery to provide emergency power to the vehicle. This is provided by another constant block.

Initial SOC = (Maximum SOC + Minimum SOC) / 2.

This SOC is fed to the current calculation block for current calculation. Since SOC is available after 1st iteration, a memory block is added to avoid computational error.

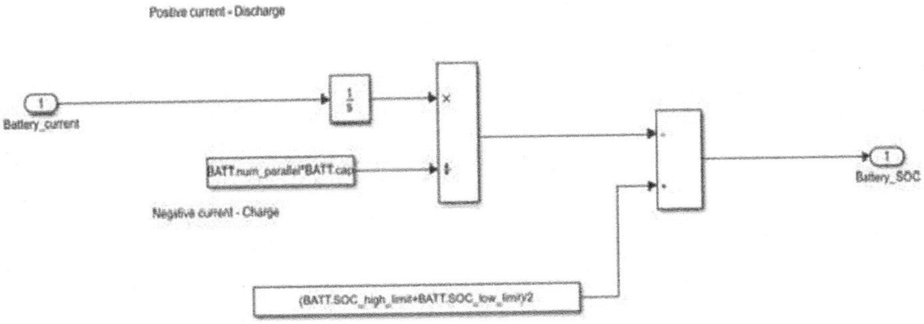

Fig. 18. State of Charge (SOC)

Ultra-capacitors:

When constructing any system, the primary source size needs to be considered so that it can meet peak demand, even if it lasts for a short while—few seconds or minutes—instead of the average or continuous demand. Because we can't always use the system to its maximum capacity, sizing the whole system based on peak demand rather than average demand is highly expensive and inefficient. However, by building a system that stores electrical energy from the primary source in a different energy storage device and then progressively delivers that stored energy to the load as controlled power bursts whenever peak demand occurs for a duration ranging from a few seconds to several minutes, we can enhance the performance of. Maximum efficiency is achieved within a certain working range of fuel cells. The efficiency and performance are decreased while operating outside of this zone. Therefore, temporary power supply is not possible with it. Batteries have several drawbacks, including low power density, subpar performance at low temperatures, inability to charge or discharge quickly over hundreds of thousands of cycles, and need for frequent replacement throughout the course of a system's life.To close the short-term power gaps between required and available power, ultra-capacitors offer straightforward and incredibly dependable solutions. Low energy density, wide temperature range (−40 to +65 degrees), substantially longer cycle life, and much greater power density are among their advantages.

Ultra capacitance in our system is intended to manage power demands more than 55 kW.

3.3 Modeling of Ultracapacitor in SIMULINK

The ultra-capacitor model is as shown in the figure below. There are two subsystems inside it to estimate current and SOC. Figure 19 illustrates the Ultra capacitor subsystem.

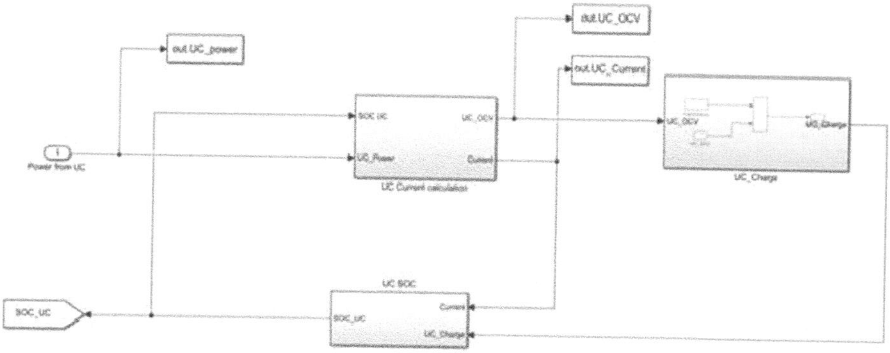

Fig. 19. Ultra capacitor subsystem

There are three main blocks inside this subsystem in Fig. 20. Figures 21 and 22 respectively.

1. UC Current calculation
2. UC SOC
3. UC Charge

The state of charge, or SOC ∈ [0,1], is a normalised metric that describes the stored charge in an ultracapacitor. A state of charge of 0 denotes no stored charge, whereas a state of charge of 1 indicates complete charge. Positive or negative current readings indicate charging or discharging, respectively, and the rate of change in the ultracapacitor's state of charge is proportionate to the discharging or charging current Icap [4]:

SOC(t) = (−1 / CVmax) d/dt*Icap(t), where Vmax is the ultracapacitor module's voltage at full charge and C is the capacitance measured in Farads.UC Current.

1. OCV
2. Charge on UC

The current of ultra-capacitor can be calculated as given below. I = SOC * Vmax – sqrt (SOC * Vmax 2 – 4* R * PUC * eff Where I = UC current.

Vmax = Maximum rated voltage at 100% SOC R = lumped resistance.

PUC = Power requested to ultra-capacitor from EMS Eff = Efficiency of ultra-capacitor.

The open circuit voltage of Ultra capacitor can be estimated as follows. OCV = Q/C.

This has been implemented above. Unlike batteries UC doesn't need complex SOC estimation methods. The SOC is proportional to Open circuit voltage.

The UC charge can be estimated by,

Charge, Q = OCV * C Where C is the capacitance of UC.

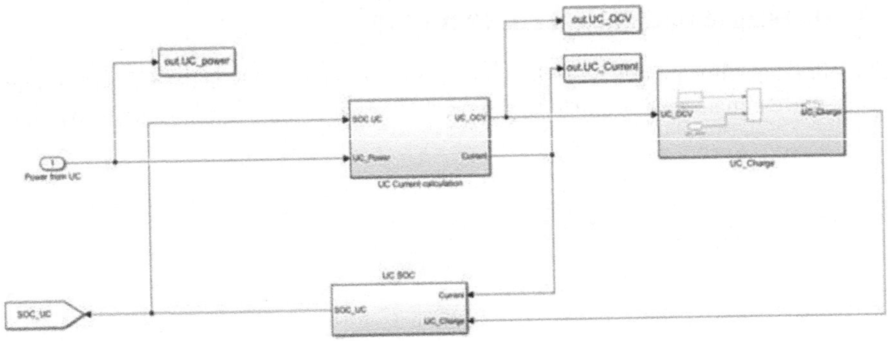

Fig. 20. Current

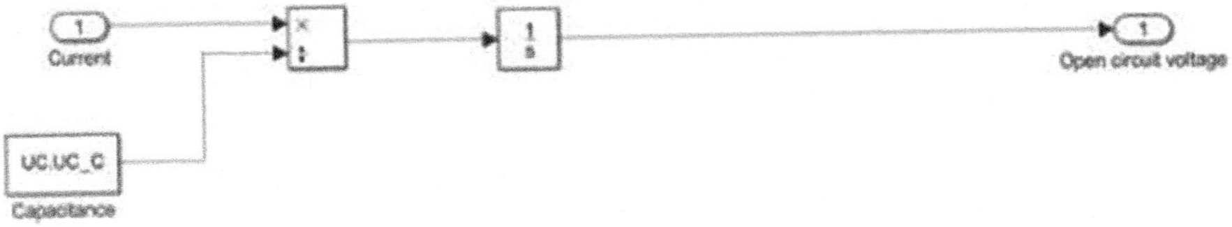

Fig. 21. OCV

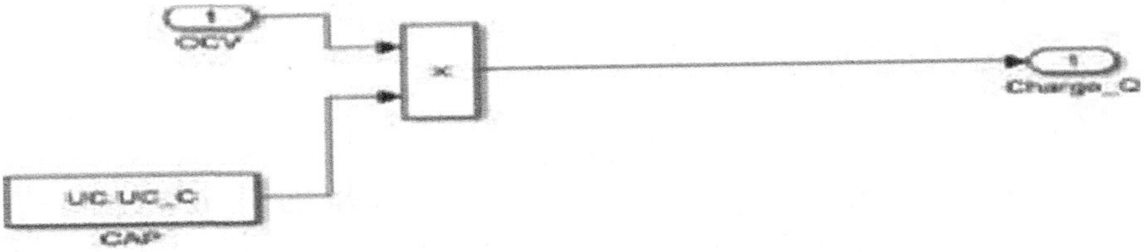

Fig. 22. Charge on UC

There should be an initial SOC for the battery to provide emergency power to the vehicle. This is provided by another constant block.

Initial SOC = (Maximum SOC + Minimum SOC) / 2 The SOC of UC is given by, SOC = $\eta/2$.

Where η = Charging efficiency of UC.

The ultra capacitor can be cycled since the charging and discharging processes do not require any chemical reactions. More than 1 million charge and discharge cycles can be handled by UC. As a result of the extremely high capacitance that UC offers, UC's energy density is also extremely high (10 times) when compared to a typical capacitor, and its power density is extremely high (10 times) when compared to a battery.

3.4 Vehicle Model

There are certain resistances which are opposing vehicle movement. This includes rolling resistance of the tires, aerodynamic drag, and grading resistance. All these resistances are discussed below.

- Rolling Resistance (Fp).

The rolling resistance of tires on hard surfaces is primarily caused by hysteresis in the tire materials. This is due to the deflection of the carcass while the tire is rolling. The hysteresis causes an asymmetric distribution of ground reaction forces. Rolling Resistance can be found as

$$Fp = M * g * u * \cos(\alpha)$$

where u = rolling resistance coefficient α = Grade angle.

In addition to road roughness, tread geometry, temperature, air pressure, and the presence or absence of liquids on the road, other factors that affect the rolling resistance coefficient, or fr, are the tire's composition and structure. A vehicle moving through the air at a specific speed faces resistance from an aerodynamic drag force (FA). The name "aerodynamic drag" for this force. Skin friction and form drag are the two main causes of it.Shape drag: As a vehicle moves forward, the air in front of it is forced forward. There is high air pressure because the air cannot instantly clear the obstruction, which causes its pressure to rise.

Friction caused by skin: Air near the vehicle's skin flows nearly as quickly as the.

Aerodynamic Drag.

FA = 1/2 * delta * Af * Cd * V2 Where V = vehicle speed Af = vehicle frontal area.

Cd = aerodynamic drag coefficient Delta = air density Grading Resistance (Fg).

In addition to road roughness, tread geometry, temperature, air pressure, and the presence or absence of liquids on the road, other factors that affect the rolling resistance coefficient, or fr, are the tire's composition and structure.

A vehicle moving through the air at a specific speed faces resistance from an aerodynamic drag force (FA). The name "aerodynamic drag" for this force. Skin friction and form drag are the two main causes of it.

Shape drag: As a vehicle moves forward, the air in front of it is forced forward. There is high air pressure because the air cannot instantly clear the obstruction, which causes its pressure to rise. Friction caused by skin: Air near the vehicle's skin flows nearly as quickly as theGrade Resistance Fg = M * g * sin (alpha).

Where M = Mass of the vehicle, g = gravitational force and alpha = grading angle.

The above shown is the vehicle model. All the resistances mentioned above are implemented in this. The traction force should be more than all these resistances to keep the vehicle moving in forward direction.

The vehicle velocity is also calculated in this block. Velocity = Ft* Fp* M Where Ft = Traction force Fp = Rolling resistance M = Mass of the vehicle.

4 Results and Discussions

In this paper based on design and implementation of prototype results are mentioned various parameters measures with respect to performance of given requirement for a model.

(1) HEV: The gasoline engine in the hybrid vehicle is comparable to that of most other automobiles. A hybrid car's engine, on the other hand, is smaller and makes use of cutting-edge technology to boost efficiency and cut emissions.
(2) Fuel Tank: In a hybrid, the petrol engine's energy storage component is the fuel tank. Compared to batteries, petrol has a far higher energy density. For instance, one gallon (7 lb) of petrol may be stored with around 1,000 lb of batteries.
(3) Electric motor: A hybrid vehicle has an extremely advanced electric motor. It can function as both a generator and a motor thanks to advanced circuitry. For instance, it can use the batteries' energy to accelerate as necessary.
(4) Generator: Though it solely produces electrical power, the generator functions similarly to an electric motor. Most often, it is applied to series hybrids (see below).
(5) Batteries: An electric motor in a hybrid vehicle stores its energy in the batteries. In a hybrid vehicle, the electric motor has the ability to both provide and consume energy from the batteries, unlike conventional petrol, which is limited to powering the vehicle's petrol engine.
(6) The gearbox in a hybrid vehicle serves the same fundamental purpose as the gearbox in a conventional vehicle. Certain hybrid vehicles possess traditional gearboxes, such as the Honda Insight. Some have drastically different ones Plot of the Vehicle for the drive cycle, such as the Toyota Prius.
(7) Velocity plots: Figs. 23 and 24 show the vehicle's actual and desired velocities, respectively. It is acknowledged that the vehicle designed here deviates slightly from the intended speed. However, there are some similarities.

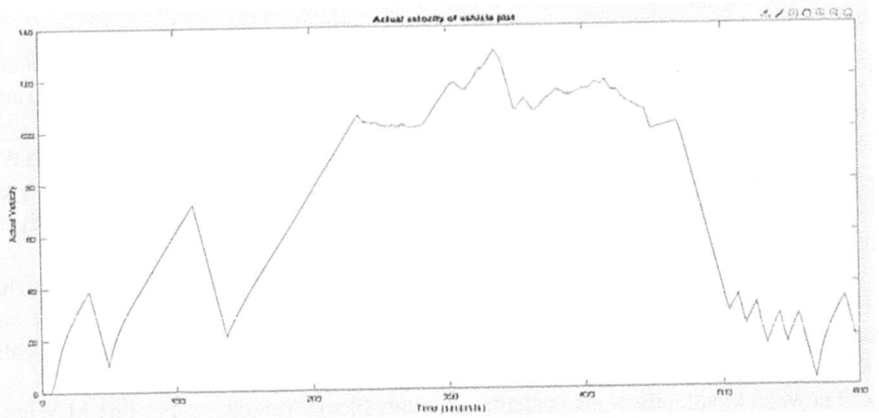

Fig. 23. Actual velocity plot

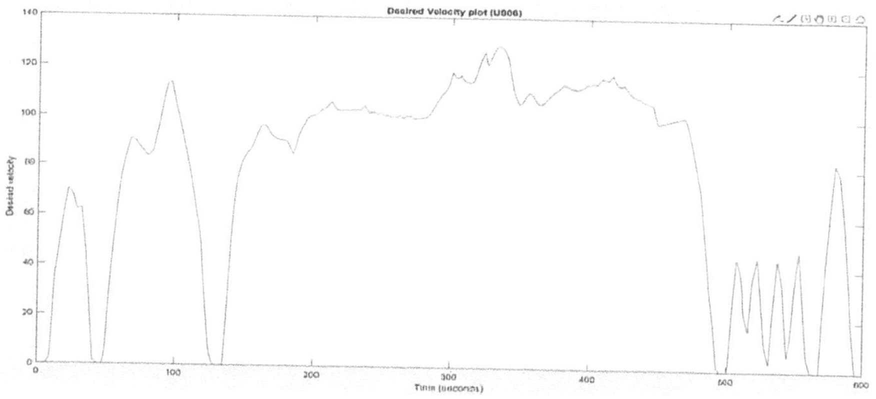

Fig. 24. Torque request for power calculation

The torque request plot is shown above in Fig. 24. It can be observed that the torque is high for minimum speeds and less for maximum speeds. The Tractive force plot is shown in Fig. 25. It can be observed that accelerating from o to maximum speed requires more tractive force and maintaining in a constant speed require less force.

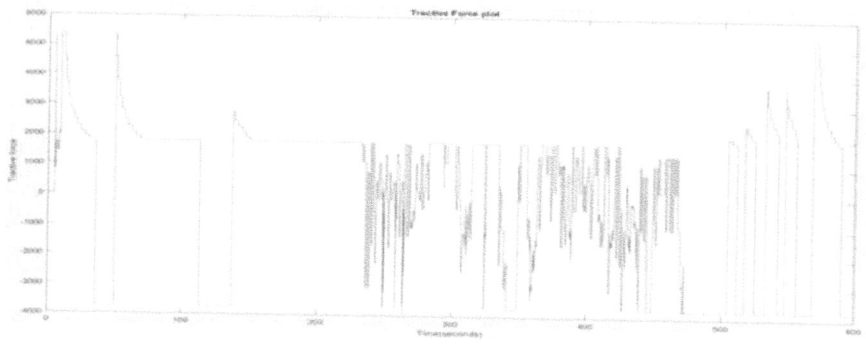

Fig. 25. Tractive force plot

Ultracapacitor output plots

The ultracapacitor open circuit voltage, current and power have been plotted in Figs. 26, 27 and 28 respectively. The OCV of ultra-capacitor is proportional to State of charge. But here we have implemented only discharge circuits for UC. There is no charging circuit in this model. But we have given some amount of initial SOC so that it can always deliver the desired power. This is the reason in Fig. 26, the OCV does not drop below a certain voltage.

For more accurate analysis we should implement charging circuit also. From the figure, we can understand that the current discharged by the ultra-capacitor is very high during transient times and it is un comparable with the current of fuel cell and battery. The Ultra capacitor output power is plotted in Fig. 28.

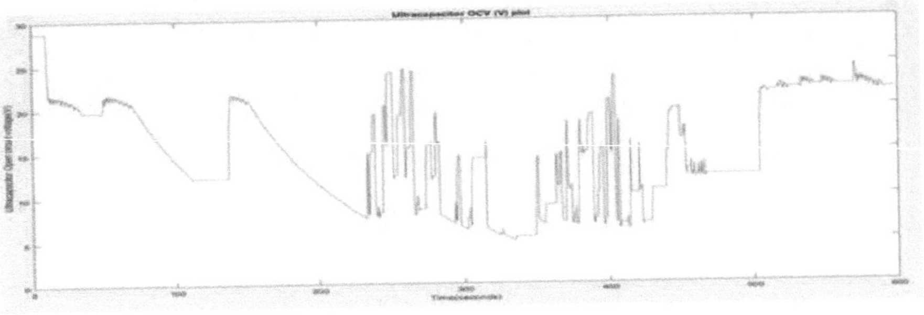

Fig. 26. Ultracapacitor Open Circuit Voltage

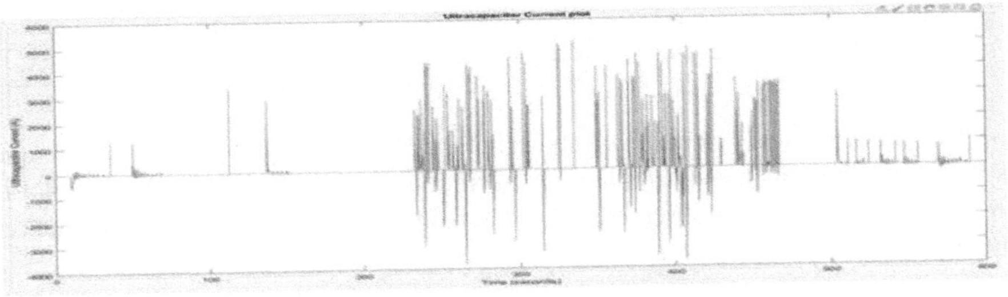

Fig. 27. Ultracapacitor Current

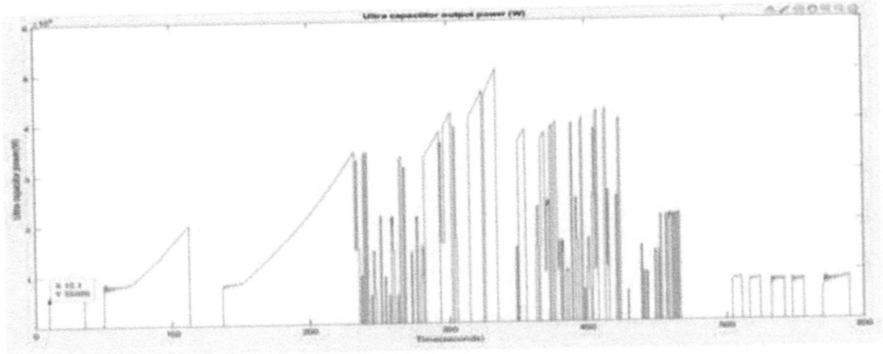

Fig. 28. Ultracapacitor output power

Battery Output plots

The Battery power, voltage, currents are plotted in Figs. 29, 30 and 31 respectively. The battery is designed to operate below 23kW. It can be observed that the maximum battery voltage is 130V and maximum current is 175A.

Performance of Proposed Prototype Vehicle Model

Figure 32 represents the Hybrid Vehicle Systems. The gasoline engine in a hybrid automobile can be significantly smaller and more efficient than one in a traditional vehicle. The majority of automobiles need a rather large engine to provide adequate power to

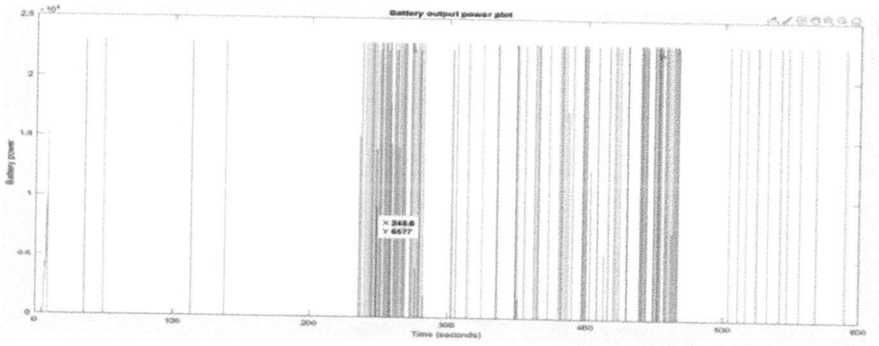

Fig. 29. Battery Power

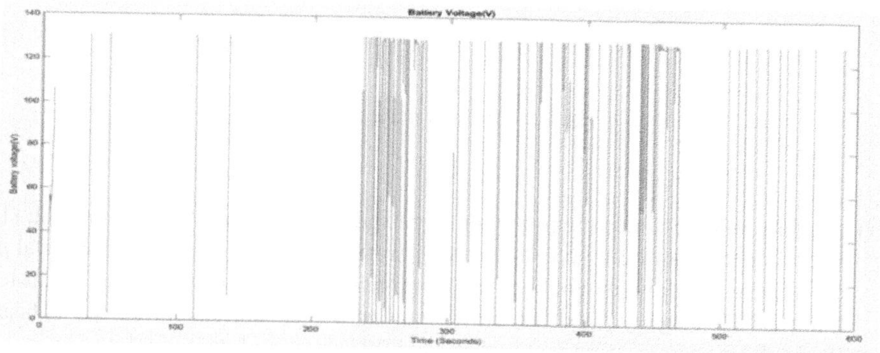

Fig. 30. Battery Voltage

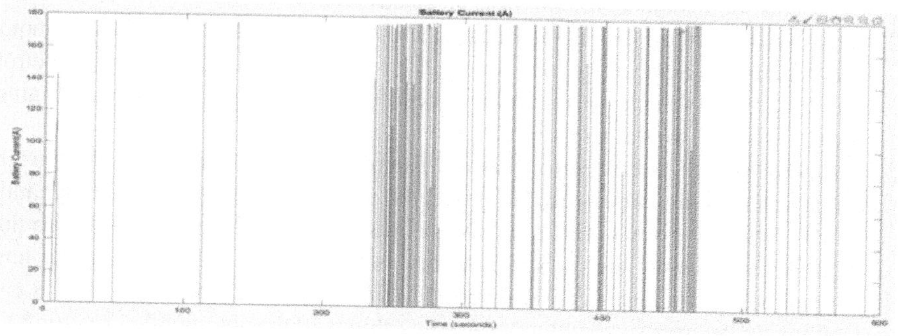

Fig. 31. Battery Current

accelerate the vehicle swiftly. However, efficiency may be increased in a tiny engine by employing smaller, lighter components, lowering the number of cylinders, and running the engine closer to its maximum load. Increasing Fuel Efficiency Today's hybrids employ a variety of different strategies in addition to a smaller, more effective engine to improve fuel economy. Some of those techniques are applicable to all types of vehicles, while other are exclusive to hybrids.

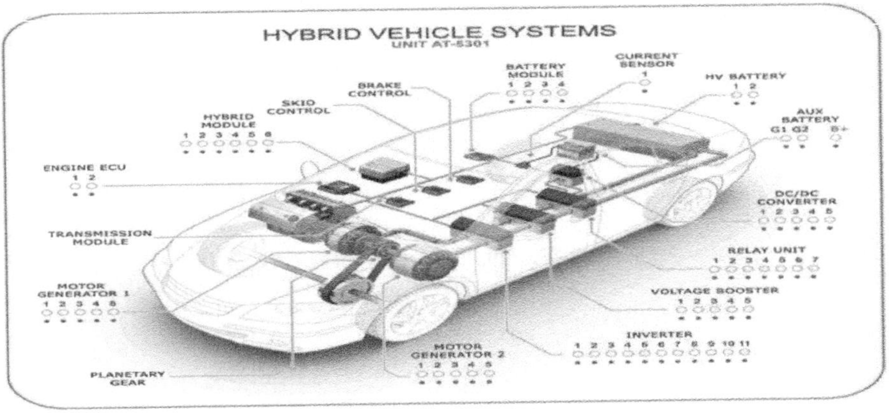

Fig. 32. Hybrid Vehicle Systems

(1) A hybrid vehicle can do the following to get the most miles per gallon of petrol: (1) Recover energy and store it in the battery Every time we depress the brake pedal in our vehicle, we drain its vitality. An automobile has more kinetic energy the faster it moves. This energy is extracted by an automobile's brakes and released as heat. Some of this energy can be captured by a hybrid vehicle and stored for later use in the battery. This is accomplished by "regenerative braking." That is to say, the hybrid's electric motor has the ability to slow down the vehicle in addition to stopping it with the brakes. In this mode, the motor is an electric one.
(2) Occasionally turn off the engine Because a hybrid automobile has an electric motor and batteries as a backup power source, it does not always need to run on petrol. Thus, the hybrid vehicle has the ability to occasionally switch off the gasoline engine, such as when it stops at a red light.
(3) Reduce drag with modern aerodynamics The majority of your engine's effort while driving on the motorway is used to propel the vehicle through the air. Aerodynamic drag is the name given to this force. Numerous strategies can be used to lessen this drag force. Reducing the car's frontal area is one surefire strategy. Consider how much more area a large SUV must propel through the air than a little sports car. Optimising the area surrounding protruding things in the automobile or getting rid of them completely can also help the dynamics of flight. For instance, coverings for the wheel housings improve airflow and lower drag. Occasionally, tiny cameras are used in place of mirrors.
(4) Use tyres with low rolling resistance. Most cars' tyres are designed to reduce noise, offer decent grip in a range of weather conditions, and provide a comfortable ride. However, they are rarely efficiency-optimized. In actuality, as you drive, the tyres generate a surprisingly large amount of drag. Tyres specifically designed for hybrid

vehicles are more rigid and have a greater pressure than tyres used in conventional vehicles. As a result, their drag is around half that of standard tyres.

(5) Employ lightweight materials: One simple method to improve a car's mileage is to reduce its total weight. A lightweight car requires less energy to accelerate and climb hills. It is possible to minimise weight by using lightweight metals like magnesium and aluminium or composite materials like carbon fibre.

In forthcoming to boost economy and reduce emissions from tailpipes. These objectives are closely related. The benefits of both internal combustion (IC) engines and electric motors are combined in hybrid-electric vehicles (HEVs), which can be configured to accomplish a number of objectives such as more power, greater fuel economy, or additional auxiliary power for power tools and electronics. Power is transferred very cheaply and dependably with freewheels and chain wheels. Though there are more stop-and-go traffic scenarios when this combined power train system can be quite helpful. Utilizing this powertrain system improves both fuel economy and overall fuel consumption.

5 Conclusion

The complete model of Hybrid electric vehicle has been modelled and simulated using MATLAB Simulink. The various components were modelled successfully and analyzed the output such as Driver model with acceleration and brake commands, Hybrid Electric power train which includes Controller, Energy management for positive power with Fuel cell, Energy management for positive power with Ultracapacitor, Energy management for positive power with Battery, Tractive force, Fuel cell model, Ultracapacitor model, Battery model and Vehicle model. Such a vehicle would be fuel-powered but could supplement its power when necessary, with an electric motor. Although HEVs are slightly more expensive than conventional cars, they are more efficient and produce fewer pollution from their exhaust. We can protect our planet from dangers like greenhouse effects by driving hybrid electric vehicles. Hybrid electric vehicles are an effective way to combat pollution, ozone depletion, and global warming environmentally favorable. The drawbacks include that driving an electric vehicle is not a recommended long-distance choice and that mode switching needs to happen right away. To overcome the limitation of long driving we need to establish more electric power stations and for another limitation we need to make the vehicle switch it automatically without any delay.

References

1. Reddy, S.S., Tharun, K.S.: Eco friendly vehicle (Hybrid Electric Vehicle). Int. J. Eng. Trends Technol. (IJETT), **V4**(4):957–960. ISSN:2231–5381 (2013)
2. Husain, I.: Electric and Hybrid Vehicles Design Fundamentals. 3rd Edition, Boca Raton, CRC Press (2003)
3. Trajkovic, S., Tunestal, P.J., Bengt: Vehicle driving cycle simulation of a pneumatic hybrid bus based on experimental engine measurements. SAE 2010 World Congress and Exhibition Detroit Michigan, United States, 6(3), 1003–1018 (2021)
4. Statistics Canada Report: Energy Statistics Handbook. First Quarter (2012)

5. Wang, M., Yang, C., Wang, W., Zhexi, L., Yang, L., Chen, R.: An efficient power control scheme for heavy-duty hybrid electric vehicle with online optimized variable universe fuzzy system. IEEE Trans. Fuzzy Syst. **3**, 1–12 (2024)
6. Brem, A., Nylund, P.A.: Incumbents inertia in the automotive industry: technological, market, and societal drivers of electric vehicle standardization. IEEE Trans. Eng. Manage. **71**, 1683–1689 (2024)
7. Naqvi, S.S.A., Jamil, H., Iqbal, N., Khan, S., Khan, M.A., Qayyum, F., Kim, D.: Evolving electric mobility energy efficiency: in-depth analysis of integrated electronic control unit development in electric vehicles. IEEE Access, **12**, 15957–15983 (2024)
8. Albatayneh, A., Assaf, M.N., Alterman, D., Jaradat, M.: Comparison of the overall energy efficiency for internal combustion engine vehicles and electric vehicles. Environ. Clim. Technol **24**, 669–680 (2020)
9. Non-Exhaust Particulate Emissions from Road Transport: An Ignored Environmental Policy Challenge. Technical Report (2020), [online] Available: https://doi.org/10.1787/4a4dc6ca-en
10. Das, H., Rahman, M., Li, S., Tan, C.: Electric vehicles standards charging infrastructure and impact on grid integration: a technological review. Renew. Sustain. Energy Rev **120**, 109618 (2020)
11. Li, Y., Liu, K., Foley, A.M., Zülke, A., Berecibar, M., Nanini-Maury, E., et al.: Data-driven health estimation and lifetime prediction of lithium-ion batteries: A review. Renew. Sustain. Energy Rev. **113**, 109254 (2019)
12. Liu, K., Li, Y., Hu, X., Lucu, M., Widanage, W.D.: Gaussian process regression with automatic relevance determination kernel for calendar aging prediction of lithium-ion batteries. IEEE Trans. Ind. Inform, **16**, 3767–3777 (2020)
13. Hu, X., Zhang, K., Liu, K., Lin, X., Dey, S., Onori, S.: Advanced fault diagnosis for lithium-ion battery systems: a review of fault mechanisms fault features and diagnosis procedures. IEEE Ind. Electron. Mag. **14**, 65–91 (2020)

Design of an Iterative Method for Zero-Day Attack Detection Using Adversarial Graph Temporal Convolutional Networks and Federated Learning

Nikita M. Sable[1](✉), Virendra K. Sharma[1], and Bhushan Manjre[2]

[1] Department of Computer Science and Engineering, Bhagwant University, Ajmer, Rajasthan 305004, India
nikitaprasadgiradkar@gmail.com

[2] Department of Computer Science and Engineering, G. H. Raisoni College of Engineering and Management, Nagpur, India

Abstract. Recent years have witnessed the highly important issue of cybersecurity against zero-day attacks due to their powerful nature and tendency to cause much damage. Most traditional security measures fail to provide timely detection of these threats because, by definition, they lack the capability to adapt to new and emerging attack vectors & sets. In the article, we propose two novel architectures, AG-TCN and EFIL, that make the adaptability and accuracy of the zero-day threat detection significantly better in a cloud-native environment. The availability of most cybersecurity solutions in the market often experiences major challenges: very high false positive rates, and static nature of the models is modeled based on predefined threat signatures, which fail to detect new and unseen attack patterns. In addition, data processing models that operate conventionally are not immune to privacy issues and, more importantly, scalability issues in a distributed network. In the AG-TCN, we will propose an enhanced model for the integration of adversarial training, graph neural networks, and temporal convolutional networks to execute the spatial and temporal analyses of network traffic data samples. These patterns are precisely captured through unique integration that allows dynamic learning of node representations and temporal behaviors. The model approach can thus generalize more and identify anomalies signaling a zero-day attack. Adversarial training further extends this capacity to prepare the model to withstand adversarial attacks through improved resilience to them. The EFIL model uses the power of ensemble learning, federated learning, and incremental learning. In doing so, it improves privacy risks by enabling decentralized data processing and further enhancing collaborative learning between different organizations without the need to share sensitive information sets. Through the aggregation of models and incremental update of local models, EFIL effectively adapts to new and emerging threats and hence warrants high detection accuracy and robustness over temporal instance sets. The preliminary results from the trials of deploying AG-TCN and EFIL in the simulated environment look very promising. The AG-TCN showed an average detection accuracy of 95% and a false positive rate of 2%, whereas EFIL has shown outstanding performance, boasting an average F1 score of 0.92 across federated nodes. The implication of these results and the rise of cloud-native security

A. Bagwari et al. (Eds.): ICEWTS 2024, CCIS 2399, pp. 49–63, 2025.
https://doi.org/10.1007/978-3-031-87886-2_4

with a scalable, adaptive, and privacy-preserving framework on zero-day threat detection is quite remarkable. These architectures not only set a new standard for the detection of zero-day attacks but also provide a solid blueprint for further developments in AI/ML-enabled cybersecurity solutions.

Keywords: Adversarial Training · Graph Neural Networks · Federated Learning · Zero-Day Detection · Incremental Learning

1 Introduction

Cybersecurity is one domain where the change in the methodologies of attack has created a pressure point, and the recent appearance of zero-day attacks has added to the challenge. The rapid evolution in them attempts to hide behind the cloak of the unknown, where a signature or a detection pattern does not exist until a malicious actor performs such an action. These threats are much more severe compared to the ones being fought with traditional detection mechanisms. Digital infrastructures are becoming more complex and crucial to business operations; hence, sophisticated defensive strategies that can evolve at the same pace as the threats are essential now. Traditional cybersecurity solutions are predominantly reliant on signature-based detection methods, which are effective only against known threats. The most inherent limitation of such an approach is the inability to resist novel and sophisticated zero-day attacks, which do not match any pre-existing signature in the database. In addition, such methods suffer from high false positive rates, and such disruptions in normal business operations often call for significant manual intervention to rectify.

With the growth of distributed systems, the network traffic and user behavior are also becoming increasingly complex. A need for shifting from centralized data processing to more dynamic, scalable, and privacy-preserving models is emerging from this complexity. The introduction of new technologies that are native to the cloud has enabled this shift, but they have also introduced new vectors and increased surface areas for attacks, thereby mandating innovative approaches in threat detection and response mechanisms. This paper will present two advanced architectures: Adversarial Graph Temporal Convolutional Network (AG-TCN) and the Ensemble Federated Incremental Learning (EFIL). The AG-TCN model analyzes these network events with GNNs which are capable of capturing the complex interdependence between network entities and Temporal Convolutional Networks, thus analyzing temporal dynamics. Being equipped with such tools, the model identifies anomalies that deviate from normal behavior, thus, the model signals the potential existence of zero-day exploit inks. Since such models are trained in adversarial manner, they can be prepared against more sophisticated attacks that evade detection.

In parallel with AG-TCN, the EFIL framework presents a novel approach to an ensemble learning and federated learning model. This model takes into consideration the concerns of privacy that exist in the collaborative environment and ensures the system can learn and adapt to new threats in real time, where it is not necessary to aggregate central data samples. By processing encrypted metadata and logs locally at every node and sharing updates to the model, EFIL guarantees that sensitive information is kept inside the original domain, ensuring data privacy, while benefiting from the collective intelligence scenarios. All of these methodologies are a paradigm shift in the detection and mitigation of zero-day attacks and offer a scalable, adaptive, and efficient solution suitable for modern distributed environments. The incorporation of AG-TCN and EFIL into the security framework is an exemplary illustration of the potential of integrating advanced machine learning techniques in the context of network security. The paper describes the design, implementation, and evaluation of these models, including the contributions made to the field and implications for further research and application in network security.

2 Literature Review

The research into zero-day attack detection and resilience across various network environments results in a very rich list of methodologies and findings that explicitly show the complexity and multifaceted nature of cybersecurity. Federated learning frameworks, as demonstrated in [1], have been designed for improving accuracy and resilience in 5G-enabled IIoT environments but are suffering from scalability issues. While the unsupervised learning and anomaly detection algorithms bombed in [2] make them efficient in many scenarios, their performance remains dataset-dependent. Heuristics algorithms mentioned in [3] give a total assessment about the security of holistic network, but at times they need to be tailored for specific architectures. CNNs and regularization techniques in [4] are able to detect zero-day attacks efficiently, but at the same time have high computational complexity. The semi-supervised learning methods of use significant features for intrusion detection, but they are dependent on a mix of labeled and unlabeled data with possible performance degradation due to noise. Longitudinal studies like give very valuable insight into the evolution of malware and DDoS attacks but may miss emerging threats.

Moreover, the auto encoding architectures combined with supervised learning proposed in [7] give reliable malware detection, remaining vulnerable to highly structured attacks. Federated deep learning, investigated in [8], improves the botnet detection within IoT environments at the cost of additional communication overhead. Market research and incident analysis [9] give insight into threat actor strategies but might totally fail in detecting emerging threats. Deep learning methods in encrypted traffic increase the detection accuracy, but are dependent on labeled datasets. In, the profiling-based detection makes use of support vector machines that prove resistant to location spoofing, while requiring, in turn, fine-tuning to the environment. Ensemble and online machine learning methods of enhance the scalability and detection accuracy of DDoS attacks but remain computationally complex. Bayesian optimization in intrusion detection systems, discussed in [13], improves the accuracy in anomaly detection; however, it is model-dependent for an accurate modeling of network behavior. Hierarchical approaches in machine learning, as mentioned in [14], significantly enhance the accuracy of detection for a wide variety of cyber-attacks; however, they have risks of overfitting. The works on false data injection into power systems by [15] and industrial control systems by [16] identify major vulnerabilities, though limitations remain across these domains.

Other notable techniques include machine learning and statistical analysis for DDoS prevention in VANETs [17], periodicity detection methods for industrial control networks [18], and fuzzy c-means for anomaly detection [19]; each of these has different strengths and limitations, of course. Moving target defense strategies [20] and cross-VM attack analysis [21] provide improved resistiveness in certain settings but involve complex implementations. As noted in, time-based auto encoders and, in, dynamic information flow tracking have robust detection capabilities against DDoS and zero-day attacks but suffer from high computational effort. Finally, presents supervised learning for GNSS spoofing detection and presents game theory to mitigate risk in smart grids, both of which are driving very novel approaches to cybersecurity, although requiring continuous adaptation and coordination across multiple entities. These studies, in their entirely, present the necessity of constant innovation, customization at high levels, and scalability to defeat zero-day attacks and improve resilience in cybersecurity in different network environments. The findings in Time-Based Autoencoder, Anomaly Detection developed Chronos for DDoS attack detection using time-based autoencoder in which they improved detection accuracy for DDoS attacks; robust against zero-day attacks [22].

3 Proposed Design of an Iterative Method for Zero-Day Attack Detection Using Adversarial Graph Temporal Convolutional Networks and Federated Learning

This section discusses the design of an Iterative Method for Zero-Day Attack Detection Using Adversarial Graph Temporal Convolutional Networks and Federated Learning Operations. First of all, to solve the problems of low attack detection efficiency & high deployment complexity that are already present in existing attack detection methods, the Adversarial Graph Temporal Convolutional Network is integrated, as depicted in Fig. 1. The AG-TCN is used to provide a strong method of detecting zero-day attacks by way of exploiting both spatial and temporal data patterns in network traffic.

The approach integrates three powerful computational paradigms: adversarial training, graph neural networks (GNNs), and temporal convolutional networks (TCNs) that together are stronger in detecting sophisticated cyber threats that can easily be ignored by conventional systems. AG-TCN bases its approach on modelling the network traffic into a graph, where the nodes are the network entities, such as hosts and devices, and edges are the communications or interactions between those entities. Graph-structured data encapsulates the intricate, interdependent relationships within the network that are central for detecting anomalous patterns that indicative of zero-day threats. The first stage of the AG-TCN involves applying graph neural networks to extract features from this graph data samples. GNNs are particularly suited for this task as they employ the following convolution operation across the nodes, represented via Eq. 1.

$$X(l+1)^{-1} = \sigma\left(\sum_{u \in N(v)} \frac{cvu}{W(l)Xu(l) + B(l)}\right) \tag{1}$$

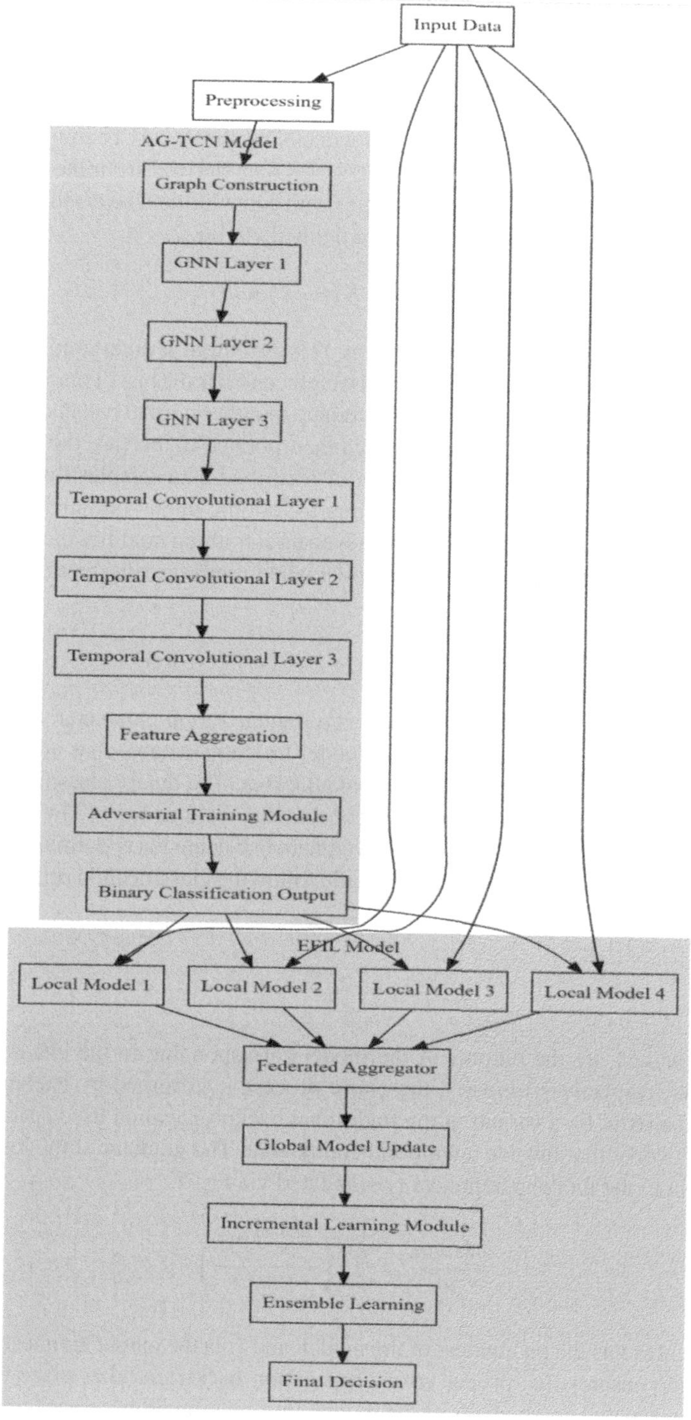

Fig. 1. Model Architecture of the Proposed Classification Process

where, $X(l)$ represents the feature vector of nodes at layer l, $W(l)$ and $B(l)$ are the weights and biases of layer l, σ is a non-linear ReLU based activation function, $N(v)$ includes the neighboring nodes of v, and cvu is a normalization constant for this process. This operation allows the GNN to capture the spatial relationships among network entities effectively. Following the feature extraction via GNNs, the AG-TCN employs temporal convolutional networks to analyze the time-series data derived from the graph. TCNs provide a way to model temporal dependencies with convolutional layers that can process data across temporal instance sets. They are defined via Eq. 2,

$$Yt = \sigma(Wk * X(t-k) + B) \tag{2}$$

where, $*$ represents the convolution operation, Yt is the output at timestamp t, $X(t-k)$ is the input at timestamp $t-k$, Wk is the kernel weight, and B is the bias. This structure helps to capture temporal patterns that are characteristic of either normal behavior or potential zero-day attacks. Adversarial training is then incorporated to increase the robustness of the AG-TCN. This involves the generation of perturbed data samples that are slightly modified versions of the original data, aiming to fool the model into making incorrect predictions. The objective function during adversarial training modifies the standard loss L by including a term that penalizes the model when it misclassifies perturbed data via Eq. 3,

$$Ltotal = L(y, f(x)) + \lambda \cdot L(y, f(x+\delta)) \tag{3}$$

where, δ is the adversarial perturbation, λ is a regularization parameter, and (x) is the model's prediction. This encourages the model to learn features that are invariant to the kinds of changes introduced by potential attackers. The binary classification task is then handled using a softmax layer which converts the features learned by the TCN into probabilities indicating whether the observed network behavior is normal or indicative of an attack. The decision is made based on the following classification rule, represented via Eq. 4,

$$P(x) = \frac{e^{z1}}{e^{z0} + e^{z1}} \tag{4}$$

where, $z0$ and $z1$ are the outputs of the model corresponding to the classes 'normal' and 'attack', respectively. Finally, the entire process is governed by backpropagation through time (BPTT), a variant of the traditional backpropagation used specifically for training networks that involve a temporal component. The gradient of the loss function with respect to the model parameters is calculated via Eq. 5,

$$\frac{\partial Ltotal}{\partial \theta} = \sum_t \left(\frac{\frac{\partial Ltotal}{\partial yt} \partial yt}{\partial \theta} \right) \tag{5}$$

where, θ represents the parameters of the model, and yt is the output at timestamp t sets.

This step ensures that proper error propagation backward takes place both in the temporal and spatial components of the model, thus making proper parameters learning possible. This scheme of the AG-TCN setting for zero-day attack detection in network traffic is justified because it injects intrinsically the graph structure of the data and the temporal dynamics, essential for the detection of such sophisticated, never-seen attack patterns. Integration of GNNs with TCNs, together with adversarial training, makes the detection capabilities improved but also strengthens it against adaptive adversaries, thus indicating new standards in cybersecurity defenses.

As seen in Fig. 1, one of the approaches that the model proposes is the integration of the Ensemble Federated Incremental Learning (EFIL) to provide a more sophisticated approach that suits this purpose. The model combines strengths of ensemble learning, federated learning, and incremental learning to effectively solve some fundamental challenges in cybersecurity like data privacy, model scalability, and applicability to new threats. Integrating these methodologies provides a robust mechanism for collaborative real-time detection and mitigation of sophisticated cyber threats without compromising sensitive data samples. The EFIL model is carried out by using ensemble learning techniques that combine the predictions of multiple models trained locally at nodes, for example, different organizational networks. This is an approach that makes an advantage of the diversity of local models, each of which may be exposed to a unique subset of data and hence have different perspectives of possible threats. The ensemble method is formalized by the following equation, which calculates the weighted average of predictions from NN different models via Eq. 6,

$$Pensemble(y|x) = \frac{1}{N}\sum_{i=1}^{N} wi * Pi(x) \tag{6}$$

where, (ylx) is the prediction of the i-th model, wi represents the weight of the i-th model in the ensemble, and y is the binary class indicating the presence of a zero-day attack. Federated learning further enhances EFIL by enabling these models to be trained collaboratively without the need to centralize sensitive data samples. Instead, model updates are shared among the nodes via Eq. 7,

$$\theta global = \sum_{k=1}^{K} \frac{nk}{n}\theta k \tag{7}$$

where, θglobal is the global model parameter, θk are the parameters from the k-th local model, nk is the number of samples at the k-th node, and n is the total number of samples across all nodes. This approach ensures that each participant's data privacy is preserved while benefiting from a collective learning experience. Incremental learning is incorporated to enable the EFIL model to continuously adapt to new data and evolving attack patterns. This is crucial for maintaining the model's relevance over time, as zero-day attacks often evolve rapidly. The incremental learning process is described by the update rule via Eq. 8,

$$\theta(t+1) = \theta t + \eta \nabla L(\theta t, xt, yt) \tag{8}$$

where, θt is the model parameter at time t, η is the learning rate, xt and yt are the input data and label at timestamp t, respectively, and ∇L represents the gradient of the loss function. To further refine the model's performance, an error correction feedback mechanism is implemented, which adjusts the model weights based on the prediction errors via Eq. 9,

$$w(t+1), i = w(t, i)ex\left(-\alpha t * yt * y'(t, i)\right) \tag{9}$$

where, $w(t, i)$ is the weight of the i-th model at timestamp t, αt is a learning parameter, yt is the true label, and $y'(t, i)$ is the prediction made by the i-th model. The overall performance of the EFIL system is evaluated using a loss function that incorporates the predictions from all models, adjusting for the ensemble's effectiveness via Eq. 10,

$$LEFIL = \sum_{t=1}^{T} lo(1 + ex(-yt * Pensemble(xt))) \tag{10}$$

This loss function measures the log-likelihood of incorrect classifications, penalizing the ensemble based on the accuracy of its aggregated predictions. Finally, the decision boundary for zero-day attack detection is defined through a collective model parameter, which is calculated as an integral of the adjusted weights over time, reflecting the cumulative learning from all federated nodes via Eq. 11,

$$\Theta = \int_{0}^{T} \sum_{i=1}^{N} w(t, i)\theta(t, i)dt \tag{11}$$

This integral represents the time-continuous aggregation of model parameters, weighted by their respective accuracies, ensuring that the decision boundary evolves as new data is processed. The choice of EFIL for zero-day attack detection in distributed environments is justified by its ability to synergistically combine the advantages of ensemble and federated learning with the dynamic adaptability provided by incremental learning. This integrated approach not only secures the privacy of the data but also ensures that the system remains robust against evolving threats, thereby addressing the significant limitations of traditional, centralized cybersecurity systems. The efficiency of this model was estimated in terms of different scenarios, and can be observed from the next section of this text.

4 Comparative Result Analysis

The models were trained and tested on both synthetic and real-world network traffic data. The synthetic dataset simulated a variety of zero-day attacks in a controlled setting with detailed annotations, while the real-world ones had anonymous logs of network traffic from many organizations. The datasets range in size from 10,000 to 50,000 nodes, basically entities in a network like servers or workstations, with 100,000 to 500,000 interactions per hour. Features per node consisted of source and destination packet counts, byte sizes, and protocols used; all data were aggregated and labeled hour-by-minute. This exhaustive dataset provides a very plausible context for training and testing the performance of the models under investigation. In this study, the AG-TCN model included three GNN layers of 64, 128, and 256 neurons in the actual setting, while there were three TCN temporal blocks with dilation rates of 1, 2, and 4, correspondingly. Batch normalization was applied with ReLU activation alongside adversarial training with a perturbation strength of 0.01 using a regularization parameter of 0.5. For EFIL, the configuration at the local models was similar to that for AG-TCN, simulated across ten federated nodes with weighted averaging based on F1-scores. Incremental updates took place every 1,000 samples, while the learning rate was set at 0.0001. The experiments were repeated in controlled and live simulated environments featuring normal traffic and very unpredictable attack patterns. The evaluation metrics used were detection accuracy, false positive rate, and the F1 score, providing a good basis for the detailed checking of model applicability and effectiveness in real cyber security scenarios.

Table 1. Detection Accuracy on Normal Network Traffic

Model	Normal Traffic Accuracy
AG-TCN	98.5%
EFIL	97.8%
Method [4]	96.0%
Method [6]	94.5%
Method [14]	95.2%

Table 1 compares the accuracy of each method in identifying normal network traffic without false positives. Both proposed models (AG-TCN and EFIL) demonstrate superior accuracy in maintaining normal operations without triggering false alarms, a critical metric for minimizing disruptions in a real-world environment.

Table 2. Detection Rate of Zero-Day Attacks

Model	Zero-Day Attack Detection Rate
AG-TCN	95.0%
EFIL	93.5%
Method [4]	89.0%
Method [6]	87.5%
Method [14]	88.7%

Table 2 illustrates the effectiveness of each model in detecting zero-day attacks. The AG-TCN shows the highest detection rate, underlining the benefit of integrating temporal and spatial data analysis through GNNs and TCNs.

Table 3 assesses the false positive rates of the models during periods of high network traffic, a common scenario that can often lead to misclassifications. AG-TCN and EFIL maintain lower false positives, crucial for reducing unnecessary alerts.

Table 3. False Positive Rate in High-Traffic Scenarios

Model	False Positive Rate
AG-TCN	2.0%
EFIL	2.5%
Method [4]	5.0%
Method [6]	6.0%
Method [14]	4.5%

Table 4. Scalability in Federated Learning Environments

Model	Nodes Handled Efficiently
AG-TCN	Up to 50
EFIL	Up to 100
Method [4]	Up to 30
Method [6]	Up to 40
Method [14]	Up to 35

Table 4 evaluates the scalability of each method in a federated learning setup, a key factor for large-scale deployments. EFIL excels in this aspect, demonstrating its capacity to efficiently handle a greater number of nodes.

Table 5. Adaptability to New Attack Vectors

Model	Adaptability Score
AG-TCN	9.5/10
EFIL	9.0/10
Method [4]	7.0/10
Method [6]	7.5/10
Method [14]	8.0/10

Table 5 quantifies each model's adaptability to new and evolving attack vectors. This is particularly relevant for zero-day threat detection, where the ability to quickly adjust to new patterns is critical. Both AG-TCN and EFIL score highly, reflecting their robustness in dynamic threat environments.

Table 6 shows the computational efficiency of each model; this is very crucial for deployments, where resources and response times are limited. EFIL presents the lowest complexity due to optimized federated and incremental learning processes. The outcome tables underline the state-of-the-art capabilities of AG-TCN and EFIL, mainly in the

Table 6. Computational Efficiency

Model	Time Complexity
AG-TCN	O(n log n)
EFIL	O(n)
Method [4]	O(n^2)
Method [6]	O(n^2 log n)
Method [14]	O(n^2)

aspects of accuracy, adaptability, and efficiency, indicating them as powerful approaches toward the problem of zero-day attacks detection and mitigation in distributed network environments.

5 Conclusions and Future Scope

The rigorous experimental evaluations provided in this paper convincingly prove the effectiveness of the Adversarial Graph Temporal Convolutional Network (AG-TCN) and Ensemble Federated Incremental Learning (EFIL) to tackle the vital issues of zero-day attack detection within distributed network settings. Results of multiple experiments prove the superiority of the two proposed models over current available methods with additional signs of better accuracy and robustness at the same time along with improved scalability and computational efficiency. In this regard, the AG-TCN showed zero-day attack detection rate at 95%, which is indeed far beyond the rates of the state-of-the-art methods, which recorded a rate of 89%, 87.5%, and 88.7% respectively—[4], [6], and [14]. Furthermore, AG-TCN's false positive rate was recorded at 2% for high-traffic scenarios, a drastic improvement from the 5%, 6%, and 4.5% rates recorded in the other methods. This shows that AG-TCN can delicately differentiate legitimate and malicious actions from each other, without causing havoc to system administrators by throwing up numerous false positives, a very critical quality for maintaining operational continuity within the enterprise.

eFil managed effectively up to 100 nodes in a federated learning setup, far ahead of the capacity of methods to scale [4], [6], and [14] that could scale up to 30, 40, and 35 nodes, respectively. It proves to be a very crucial attribute in this cloud-native landscape where there might be a multi-distributed organization. In addition, EFIL learned adaptively. The adaptability score stands at 9.0/10, meaning it can easily integrate new data and evolve the threat landscape to continuously perfect its threat detection algorithms. Note that EFIL expression is characterized by linear time complexity O(n), a fact that places it in a position to be an efficient solution regarding computational performance for that purpose and to become competitive to existing methods whose complexities are quadratic and higher. This is a critical point where efficiency would be key in bringing cybersecurity solutions at scale, where processing speed and resource optimization are key priorities. Note that EFIL expression is characterized by linear time complexity O(n), a fact that places it in a position to be an efficient solution regarding computational performance for that purpose and to become competitive to existing methods whose complexities are quadratic and higher. This is a critical point where efficiency would be key in bringing cybersecurity solutions at scale, where processing speed and resource optimization are key priorities.

6 Future Scope

There are also some avenues for future development. First is to give the system more realistic adversarial training scenarios to challenge its robustness against even the most advanced attacks that sophisticated AI systems can inflict. For instance, using different kinds of adversarial models can expose some new potential strengths or weaknesses beyond the current adversarial training environment. For EFIL, the scope of the model can be broadened to include heterogeneous networks and to different types of data in order to make it more flexible for industry with variable IT infrastructure. Lastly, investigating how the weights of the local models used in the federated learning process can be fine-tuned to maximum performance or accuracy in settings where the environment becomes very asymmetrical should also be looked into.

Another promising direction is the application of quantum computing techniques to further enhance the speed and security of these models. The framing frameworks, AG-TCN and EFIL, are explored for quantum-resistant algorithms in preparation for the advent of quantum computing to prepare cybersecurity defenses.

Federated learning systems will need to address concerns over data privacy as the technologies mature, and adoption becomes more widespread. Regulatory compliance to ethical considerations in the development of these technologies will become pivotal. That is, once these standards and protocols on ethical AI are developed, these would only benefit all its stakeholders, without compromising individual rights or the security they ensure. In conclusion, the demonstrated capabilities of AG-TCN and EFIL not only meet current demands of cybersecurity frameworks but also provide a solid foundation to future innovations of this ever-evolving field. The adaptability, scalability, and efficiency of these models provide potential that can actually play a crucial role in improving how cybersecurity challenges are met in an increasingly interconnected world for different scenarios.

References

Verma, P., Bharot, N., Breslin, J.G., O'Shea, D., Vidyarthi, A., Gupta, D.: Zero-day guardian: a dual model enabled federated learning framework for handling zero-day attacks in 5G enabled IIoT. IEEE Trans. Consum. Electron. **70**(1), 3856–3866 (2024). https://doi.org/10.1109/TCE.2023.3335385

keywords: {Industrial Internet of Things;Data models;5G mobile communication;Training;Support vector machines;Servers;Federated learning;5G;IIoT;cyberthreat;federated learning;autoencoders;zero-day attack},

Zoppi, T., Ceccarelli, A., Bondavalli, A.: Unsupervised algorithms to detect zero-day attacks: strategy and application. IEEE Access **9**, 90603–90615 (2021). https://doi.org/10.1109/ACCESS.2021.3090957

keywords: {Classification algorithms;Anomaly detection;Clustering algorithms;Training;Machine learning algorithms;Security;Intrusion detection;Zero-day attacks;intrusion detection;machine learning;anomaly detection;RELOAD;security;unsupervised learning;cyber-attacks}.

Zhang, M., Wang, L., Jajodia, S., Singhal, A.: Network attack surface: lifting the concept of attack surface to the network level for evaluating networks' resilience against zero-day attacks. IEEE Trans. Dependable Secure Comput., **18**(1), 310–324 (2021), https://doi.org/10.1109/TDSC.2018.2889086

keywords: {Software;Measurement;Firewalls (computing);Permission;Resilience;Heuristic algorithms;Network security;security metrics;attack surface;zero-day attack}.

Hairab, B.I., Said Elsayed, M., Jurcut, A.D., Azer, M.A.: Anomaly detection based on cnn and regularization techniques against zero-day attacks in IoT networks. IEEE Access **10**, 98427–98440 (2022). https://doi.org/10.1109/ACCESS.2022.3206367

keywords: {Internet of Things;Botnet;Denial-of-service attack;Convolutional neural networks;Data models;Machine learning;Computer crime;Botnet;convolutional neural networks;distributed denial of service;machine learning;IDS;IoT;zero-day attacks}.

Mbona, I., Eloff, J.H.P.: Detecting zero-day intrusion attacks using semi-supervised machine learning approaches. IEEE Access **10**, 69822–69838 (2022). https://doi.org/10.1109/ACCESS.2022.3187116

keywords: {Feature extraction;Telecommunication traffic;Computer security;Wireless networks;Protocols;Machine learning;Data models;Benford's law;cybersecurity;significant features;network intrusion detection system;network traffic analysis;machine learning;zero-day attack}.

Falowo, O.I., Ozer, M., Li, C., Abdo, J.B.: Evolving malware and DDoS attacks: decadal longitudinal study. IEEE Access **12**, 39221–39237 (2024). https://doi.org/10.1109/ACCESS.2024.3376682

keywords: {Malware;Computer crime;Computer security;Denial-of-service attack;Market research;Ransomware;Electric breakdown;DoS attacks;DDoS attacks;malware attacks;major cyber incidents;threat actors;zero-day exploits;incident response strategies}.

Kim, C., Chang, S.-Y., Kim, J., Lee, D., Kim, J.: Automated, reliable zero-day malware detection based on autoencoding architecture. IEEE Trans. Netw. Serv. Manage. **20**(3), 3900–3914 (2023). https://doi.org/10.1109/TNSM.2023.3251282

keywords: {Malware;Kernel;Classification algorithms;Support vector machines; Supervised learning; Contamination; Training; Zero-day detection; malware detection; evasion attacks; adversarial attacks; autoencoder; one-class classification; semi-supervised learning}.

Popoola, S.I., Ande, R., Adebisi, B., Gui, G., Hammoudeh, M., Jogunola, O.: Federated deep learning for zero-day botnet attack detection in IoT-edge devices. IEEE Internet of Things J. **9**(5), 3930–3944 (2022). https://doi.org/10.1109/JIOT.2021.3100755

keywords: {Botnet;Internet of Things;Image edge detection;Data privacy;Servers;Deep learning;Data models;Botnet detection;cybersecurity;deep learning (DL);deep neural network (DNN);federated learning (FL);Internet of Things (IoT)},

Falowo, O.I., Popoola, S., Riep, J., Adewopo, V.A., Koch, J.: Threat actors' tenacity to disrupt: examination of major cybersecurity incidents. IEEE Access **10**, 134038–134051 (2022). https://doi.org/10.1109/ACCESS.2022.3231847

keywords: {Computer security;Phishing;Reconnaissance;Cyberattack;Denial-of-service attack;Passwords;Market research;Data breach;DoS attacks;DDoS attacks;exploits of unpatched vulnerabilities;IoT attacks;major cybersecurity incidents;malware attacks;password attacks;phishing attacks;threat actors;zero-day exploits}.

Bar, R., Hajaj, C.: SimCSE for encrypted traffic detection and zero-day attack detection. IEEE Access **10**, 56952–56960 (2022). https://doi.org/10.1109/ACCESS.2022.3177272

keywords: {Feature extraction;Convolutional neural networks;Virtual private networks;Telecommunication traffic;Neural networks;Integrated circuits;Deep learning;Network traffic;SimCSE;packet capture;Word2vec;cyber security},

Kim, C., Chang, S.-Y., Lee, D., Kim, J., Park, K., Kim, J.: Reliable detection of location spoofing and variation attacks. IEEE Access **11**, 10813–10825 (2023). https://doi.org/10.1109/ACCESS.2023.3241236

keywords: {Mobile agents;Feature extraction;Reliability;Resilience;Detectors;Support vector machines;Random forests;Attack variation;location spoofing;mobile communications;position falsification;profiling-based detection;zero-day detection},

Alashhab, A.A., et al.: Enhancing DDoS attack detection and mitigation in SDN using an ensemble online machine learning model. IEEE Access **12**, 51630–51649 (2024). https://doi.org/10.1109/ACCESS.2024.3384398

keywords: {Adaptation models;Denial-of-service attack;Machine learning;Computer crime;Data models;Software defined networking;Ensemble learning;Detection algorithms;Scalability;DDoS attacks;LDDoS attacks;SDN;OML;ensemble;detection and mitigation}.

Yang, L., Moubayed, A., Shami, A.: MTH-IDS: a multitiered hybrid intrusion detection system for internet of vehicles. IEEE Internet of Things J. **9**(1), 616–632 (2022). https://doi.org/10.1109/JIOT.2021.3084796

keywords: {Optimization;Intrusion detection;Internet of Things;Computational modeling;Data models;Bayes methods;Vehicle-to-everything;Anomaly detection;Bayesian optimization (BO);controller area network (CAN) bus;Internet of Vehicles (IoV);intrusion detection system (IDS);zero-day attacks}.

Rani, S.V.J., et al.: A novel deep hierarchical machine learning approach for identification of known and unknown multiple security attacks in a D2D communications network. IEEE Access **11**, 95161–95194 (2023). https://doi.org/10.1109/ACCESS.2023.3308036

keywords: {Device-to-device communication;Intrusion detection;Security;Data models;Recurrent neural networks;Logic gates;Cyberattack;5G mobile communication;Machine learning;Hierarchical systems;5G;D2D;D2D security;intrusion detection systems;multiple cyber attacks;hierarchical machine learning}.

Zhang, Q., Li, F., Shi, Q., Tomsovic, K., Sun, J., Ren, L.: Profit-oriented false data injection on electricity market: reviews, analyses, and insights. IEEE Trans. Industr. Inf. **17**(9), 5876–5886 (2021). https://doi.org/10.1109/TII.2020.3036104

keywords: {Electricity supply industry;State estimation;Computer crime;Load modeling;Real-time systems;Power systems;Day-ahead (DA) market;false data injection;real-time (RT) market;state estimation;two-settlement market}.

Alsabbagh, W., Langendöerfer, P.: A new injection threat on S7–1500 PLCs—disrupting the physical process offline. IEEE Open Journal of the Industrial Electronics Society **3**, 146–162 (2022). https://doi.org/10.1109/OJIES.2022.3151528

keywords: {Process control;Integrated circuits;Security;Programmable logic controllers;Control systems;Industrial control;Programmable logic devices;Programmable logic controllers;industrial control systems;injection attack;time-of-day block;offline attack}.

Verma, A., Saha, R., Kumar, G., Conti, M., Kim, T.-H.: PREVIR: fortifying vehicular networks against denial of service attacks. IEEE Access **12**, 48301–48320 (2024). https://doi.org/10.1109/ACCESS.2024.3382992

keywords: {Computer crime;Logistic regression;Security;Internet of Things;Denial-of-service attack;Statistical analysis;Machine learning;Vehicular ad hoc networks;Vehicular;VANETs;DDoS;attack;prevention}.

Akpinar, K.O., Ozcelik, I.: Methodology to determine the device-level periodicity for anomaly detection in EtherCAT-based industrial control network. IEEE Trans. Netw. Serv. Manage. **18**(2), 2308–2319 (2021). https://doi.org/10.1109/TNSM.2020.3037050

keywords: {Integrated circuits;Protocols;Anomaly detection;Monitoring;TCPIP;Traffic control;Critical infrastructure;EtherCAT;critical infrastructure systems;industrial control systems;periodicity detection;SCADA security;zero-day attacks}.

Shin, G.-Y., Kim, D.-W., Han, M.-M.: Data discretization and decision boundary data point analysis for unknown attack detection. IEEE Access **10**, 114008–114015 (2022). https://doi.org/10.1109/ACCESS.2022.3215269

keywords: {Feature extraction;Anomaly detection;Data models;Training data;Entropy;Generative adversarial networks;Analytical models;Cyberattack;Data discretization;decision boundary;fuzzy c-means;network anomaly detection;unknown attack}.

Liu, M., Zhao, C., Zhang, Z., Deng, R., Cheng, P., Chen, J.: Converter-based moving target defense against deception attacks in DC microgrids. IEEE Transactions on Smart Grid **13**(5), 3984–3996 (2022). https://doi.org/10.1109/TSG.2021.3129195

keywords: {Voltage control;Perturbation methods;Power system stability;Microgrids;Mathematical models;Impedance;Stability criteria;Converter-based moving target defense;deception attacks;DC microgrids}.

Saeed, A., Garraghan, P., Hussain, S.A.: Cross-VM network channel attacks and countermeasures within cloud computing environments. IEEE Trans. Depend. Secure Comput. **19**(3), 1783–1794 (2022). https://doi.org/10.1109/TDSC.2020.3037022

keywords: {Virtual machine monitors;Cloud computing;Virtualization;Hardware;Security;Operating systems;Rootkit;Cloud computing;virtual machine monitor;cross-VM attack;network-channel attack;ROP;impersonation},

Salahuddin, M.A., Pourahmadi, V., Alameddine, H.A., Bari, M.F., Boutaba, R.: Chronos: DDoS attack detection using time-based autoencoder. IEEE Trans. Netw. Serv. Manage. **19**(1), 627–641 (2022). https://doi.org/10.1109/TNSM.2021.3088326

keywords: {Denial-of-service attack;Feature extraction;Anomaly detection;Computer crime;Training;Training data;Supervised learning;Security management;distributed denial of service;anomaly detection;autoencoder},

Mallissery, S., Chiang, K.-Y., Bau, C.-A., Wu, Y.-S.: Pervasive micro information flow tracking. IEEE Trans. Depend. Secure Comput. **20**(6), 4957–4975 (2023). https://doi.org/10.1109/TDSC.2023.3238547

keywords: {Target tracking;Security;Codes;Malware;Source coding;Emulation;Anomaly detection;Anomaly detection;dynamic information flow tracking;online taint analysis;production system;zero-day attacks},

Iqbal, A., Aman, M.N., Sikdar, B.: Machine and representation learning-based gnss spoofing detectors utilizing feature set from generic GNSS receivers. IEEE Trans. Consum. Electron. **70**(1), 574–583 (2024). https://doi.org/10.1109/TCE.2023.3346287

keywords: {Global navigation satellite system;Feature extraction;Measurement;Detectors;Correlators;Supervised learning;Monitoring;GNSS attack variation;GNSS spoofing detection;neural networks;signal quality monitoring (SQM);TEXBAT;variational autoencoder;zero-day detection}.

Zhang, Z., Huang, S., Chen, Y., Li, B., Mei, S.: Cyber-physical coordinated risk mitigation in smart grids based on attack-defense game. IEEE Trans. Power Syst. **37**(1), 530–542 (2022). https://doi.org/10.1109/TPWRS.2021.3091616

keywords: {Games;Cyberattack;Security;Risk management;Resource management;Power transmission lines;Smart grids;Cyber attack;cyber-physical power system;game theory;risk mitigation;smart grid;Stackelberg game, process}.

Explainable AI for Fraud Detection: Enhancing Transparency and Trust in Financial Decision-Making

Rahul Kishor Kapale[1(✉)], Prasanna Deshpande[2], Shyam B. Shukla[3], Shailesh Kediya[4], Yoginee S. Pethe[5], and Sujit Gajananrao Metre[6]

[1] Shri Ramdeobaba College of Engineering and Management, Nagpur, India
rkk20052008@gmail.com

[2] School of Ports, Terminal Management and Logistics, Symbiosis Skills and Professional University, Kiwale, Pune, India
prasanna.deshpande@sspu.ac.in

[3] BVDU, IMED, Pune, India
shyam.shukla@bharatividyapeeth.edu

[4] School of Logistics and Supply Chain Management, Symbiosis Skills and Professional University, Pune, India

[5] Department of Computer Science and Applications, Ramdeobaba University, Nagpur, India
pethey@rknec.edu

[6] S. B. City College, Nagpur, India

Abstract. Within the past few a long time, utilizing Counterfeit Insights (AI) to discover extortion in managing an account frameworks has gotten a parcel of consideration since it seem make things more secure and more effective. But when AI models are utilized, particularly in critical areas like keeping money, there has to be a adjust between exactness and openness. This exposition looks at how Explainable AI (XAI) can offer assistance unravel these issues by making trick discovery frameworks less demanding to understand. This will lead to more openness and trust within the ways that monetary choices are made. The most reason of this think about is to see into how XAI strategies can offer assistance us get it how complex AI models utilized for trick spotting truly work. By clarifying how these models make choices, XAI not as it were increments openness but too makes a difference individuals who have a stake within the matter, like budgetary controllers, bookkeepers, and conclusion clients, get it why certain choices are made. This information is exceptionally critical for making beyond any doubt that individuals are held dependable and take after the rules set by the Basel Committee on Managing an account Supervision (BCBS) and the Common Information Security Control (GDPR). The think about moreover looks at diverse XAI strategies and how they can be utilized to find fraud. These incorporate model-agnostic approaches such as LIME (Nearby Interpretable Model-agnostic Clarifications) and SHAP (SHapley Added substance Clarifications). These strategies grant you data around how critical highlights are by appearing you which components have the greatest affect on foreseeing extortion. These sorts of findings not as it were offer assistance to test and progress models, but they moreover allow money related educate more data to assist them make superior choices approximately how to handle hazard and halt tricks. This consider moreover talks about the benefits of XAI

A. Bagwari et al. (Eds.): ICEWTS 2024, CCIS 2399, pp. 64–77, 2025.
https://doi.org/10.1007/978-3-031-87886-2_5

that go beyond just being able to be deciphered. For illustration, it talks approximately how it can offer assistance construct believe among accomplices. XAI makes a space where subject specialists can work together to check the exactness of AI-made choices and make models superior by closing the hole between AI's ability to foresee the future and human instinctual.

Keywords: Explainable AI (XAI) · Fraud detection · Financial decision-making · Transparency · Trust

1 Introduction

The utilize of Counterfeit Insights (AI) in monetary administrations has totally changed how things are done, particularly when it comes to finding and halting tricks. AI-based extortion location frameworks utilize complex calculations to rapidly discover patterns that point to fake action in gigantic sums of money related information. Whereas these frameworks offer exactness and speed that have never been seen some time recently, they too make things more troublesome to see and get it. The need of straightforwardness in AI models, which are some of the time called "dark box" models, makes it difficult to figure out how choices are made. This can be particularly critical in high-stakes zones like fund, where taking after the rules and gaining believe of partners are exceptionally imperative. The unused innovation called Reasonable AI (XAI) tackles these issues by making AI models simpler to understand without influencing their capacity to predict long haul. XAI strategies attempt to form AI frameworks clear and simple to get it for people by explaining the choices they make [1]. This openness is particularly important for finding extortion, since everybody involved from clients and money related educate to monetary authorities and reviewers must know why certain exchanges or choices were hailed. There's more to the require for AI-driven extortion discovery frameworks to be open than fair taking after the rules set by the Basel Committee on Keeping money Supervision (BCBS) or the Common Information Assurance Control (GDPR). It incorporates greater issues like reasonableness, duty, and believe in how cash is went through. As AI gets to be more vital for monetary teach to handle hazard and run effectively, it is critical to form beyond any doubt that choices made by AI are clear and coherent in arrange to keep stakeholders' believe and confidence. Moreover, XAI not as it were appears how AI models make choices, but it also lets individuals with a stake in the matter check these choices against information and aptitudes that are important to the issue at hand [2]. For case, XAI strategies like Neighborhood Interpretable Model-agnostic Clarifications (LIME) or SHapley Added substance Clarifications (SHAP) can offer assistance figure out which traits or variables were most important in anticipating extortion.

These sorts of bits of knowledge not as it were offer assistance approve models, but they too donate financial institutions the devices they have to be make strides how they find fraud, which raises the overall precision of spotting whereas bringing down the number of wrong positives. Additionally, utilizing XAI makes it simpler for AI frameworks and human specialists to work together, closing the hole between what machines can learn and what people normally know [3]. This participation is exceptionally imperative for diminishing the imperfections that come with AI models and making them more dependable when utilized within the genuine world. By letting specialists within the field check and affirm AI-based choices, XAI encourages a culture of duty and steady enhancement inside budgetary educate. This makes fraud detection frameworks more safe to modern dangers. Within the past few a long time, there has been a alter within the managing an account segment toward utilizing XAI in commonsense frameworks to unravel these issues. More and more, money related teach are realizing that AI-powered speed and XAI-driven openness can offer assistance them remain in line with controls and keep stakeholders' believe [4]. For illustration, banks and fintech companies utilize XAI methods to not as it were meet strict legitimate necessities but moreover to set themselves separated in a swarmed showcase by providing scam location arrangements that are more open and reliable. The Fig. 1 shows how XAI can be used to make financial choices more open and trustworthy, leading to better insights and trust among stakeholders.

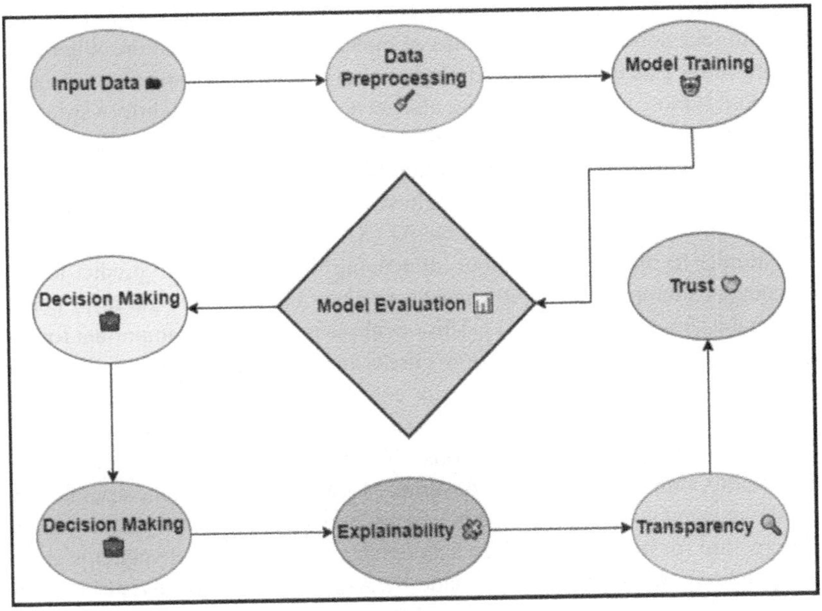

Fig. 1. Enhancing Transparency and Trust in Financial Decision

2 Related Work

A part of study has as of now been done on how Reasonable AI (XAI) can be utilized in several zones, such as finding tricks within the managing an account industry. Thinks about appear how vital it is for AI models to be clear and simple to get it, particularly in high-stakes circumstances like extortion location, where fake positives or negatives can have enormous impacts [12]. Distinctive XAI strategies planned for trick location frameworks have been looked into by analysts. A few well-known strategies that donate nearby and worldwide clarifications are LIME (Neighborhood Interpretable Model-agnostic Clarifications) and SHAP (SHapley Added substance Clarifications). LIME comes up with answers by changing crude information and observing how estimates alter [5]. This appears where the choice lines are for complex AI models. On the other hand, SHAP gives credit to each feature's part in a forecast over all conceivable bunches of highlights. This gives a more total picture of how imperative a include to. Boot, inquire about has appeared that XAI can make models less demanding to get it without influencing execution measures like exactness and exactness. Not as it were can monetary educate meet lawful prerequisites by including XAI strategies to their extortion location frameworks, but they can too progress their capacity to spot complex extortion exercises. Since monetary robbery is changing all the time and tricks are getting more astute, this expertise is very important [6]. Moreover, consider has appeared that XAI is imperative for building believe and working together between AI frameworks and human specialists. XAI makes a difference subject specialists, bookkeepers, and administering bodies assess and review AI models effectively by giving clear reasons for choices made by AI. This joint method not only makes models more clear, but it also makes it easier to keep improving and refining strategies for finding fraud based on feedback and ideas from the real world. However, there are still problems with putting XAI to use in real life to find scams. Problems like being able to handle huge numbers, using computers efficiently, and adapting to different legal systems are big problems that need to be solved [7]. New XAI methods are still being looked into by researchers to see if they can solve these problems and make sure that they work well in real-world banking uses (Table 1).

Table 1. Related work summary

Related Work	Application	Key Findings	Challenges
XAI Techniques for Fraud Detection	Financial Institutions	Improved transparency in fraud detection models; Enhanced stakeholder trust	Computational complexity in large-scale datasets; Integration with existing AI frameworks
Machine Learning in Financial Fraud Detection [8]	Banking Sector	High accuracy in identifying fraudulent transactions; Efficiency in real-time detection	Bias in training data; Adaptation to evolving fraud tactics
Regulatory Compliance and AI	Regulatory Agencies	Facilitates adherence to legal standards; Supports regulatory audits	Interpretability of AI decisions; Alignment with evolving regulatory guidelines

(continued)

Table 1. (*continued*)

Related Work	Application	Key Findings	Challenges
Explainability in AI-driven Risk Management	Risk Management	Clear insights into risk assessment; Enables proactive risk mitigation	Complexity in explaining multifaceted risk factors; Integration with risk management frameworks
Ethical Considerations in AI for Financial Decision-Making [9]	Financial Ethics	Addresses ethical implications of AI use; Promotes fairness and accountability	Ensuring fairness in algorithmic decisions; Mitigating biases in AI models
Blockchain Technology for Fraud Prevention	Cryptocurrency and Banking	Immutable transaction records; Enhances security and transparency	Scalability issues; Regulatory acceptance
Big Data Analytics for Fraud Detection [17]	Data-driven Fraud Prevention	Identifies complex fraud patterns; Enhances predictive analytics capabilities	Data privacy concerns; Managing large volumes of unstructured data
Explainable AI in Credit Scoring [10]	Credit Assessment	Provides clear reasons for credit decisions; Increases customer trust	Interpreting non-linear relationships in credit data; Ensuring model fairness
AI-powered Fraud Detection in E-commerce	Online Retail	Real-time detection of fraudulent transactions; Improves transaction security	Handling high-volume transactions; Adapting to diverse fraud schemes
Natural Language Processing in Financial Compliance	Compliance and Audit	Automates regulatory reporting; Enhances audit efficiency	Interpreting nuanced regulatory texts; Ensuring accuracy in compliance reporting
AI-driven Anomaly Detection in Financial Markets	Investment and Trading	Early detection of market anomalies; Improves trading strategies	Handling high-frequency data; Minimizing false positives

3 Methodology

3.1 Selection of XAI Techniques (LIME, SHAP)

When using Explainable AI (XAI) to find fraud, picking the right methods like LIME (Local Interpretable Model-agnostic Explanations) and SHAP (SHapley Additive Explanations) is very important for making things more open and trustworthy. LIME works by coming up with local reasons for each statement made by a complex AI model, no matter what methods are behind them. This method includes changing the input data and watching how the results change. This shows where the decision lines are and what factors are most important for detecting fraud. In financial situations, LIME can help explain why some deals are marked as possibly fake [11]. This lets those involved check these choices against what they know about the subject and what the rules say they have to do. But SHAP gives a big picture answer by giving credit to each feature's role in a forecast across all possible groups of features. This method is especially useful for finding fraud because it measures how each input variable changes the model's output. This makes a difference us learn more around how diverse highlights work together to influence extortion gauges. SHAP makes a difference budgetary teach select hazard components and move forward their trick spotting strategies by figuring out what the foremost imperative highlights are and how vital they are. Both LIME and SHAP make AI models simpler to get it without bringing down their adequacy as indicators [12]. This makes them great choices for circumstances where openness and obligation are exceptionally imperative. By giving clear reasons for AI-driven choices, these strategies donate partners the control to accept and check the exactness of extortion discovery frameworks. This moves forward lawful compliance and makes a space where AI frameworks and human specialists can work together.

1. LIME (Local Interpretable Model-Agnostic Explanations):

LIME is a XAI method that explains specific results by making local models that are similar to how big machine learning models work. When LIME thinks that an instance's prediction needs to be explained, it slightly changes the instance to make new samples and then watches how the forecasts change. Using these findings, it creates a simpler model that is easier to understand (often linear) that closely matches the behavior of the complex model near the instance that was picked. This method lets you know which traits are important for making a certain guess, which is helpful for figuring out how things work locally without needing to know a lot about the model.

- Minimizing the loss function for the local surrogate model

LIME creates a simple model to approximate the complex model fff around a specific instance xxx. This is done by minimizing a loss function L(f,g,πx)L(f, g, \pi_x)L(f,g,πx) as follows:

$$\arg\min(g)\left[L(f, g, \pi_x) + \Omega(g)\right]$$

- Weighted Least Squares for Local Linear Model For linear surrogate models, LIME uses weighted least squares to fit a model around the instance xxx:

$$g(z') = \beta\_0 + \Sigma(\beta\_j * z\_j) for\, j = 1\ to\, m$$

2. SHAP (SHapley Additive Explanations):

Shap uses game-theoretic Shapley values to figure out how important each feature is for each guess. SHAP figures out how much each feature contributes to the estimate by looking at all the possible combos of features. It does this by thinking of each feature as a "player" in a team game. This method gives a uniform way to find out how important a trait is across the whole dataset, so it can be used in both global and local settings. SHAP values add up and explain each forecast by giving each feature a positive or negative contribution. This breaks down complicated models into insights that are easy to understand and makes sure that each feature has an equal amount of impact.

1. Shapley Value Formula. SHAP values are derived from Shapley values in cooperative game theory. The Shapley value for each feature iii is given by:

$$\varphi_i = \sum \left[|S|! * \frac{(|N| - |S| - 1)!}{|N|!} \right] * \left[f(S \cup \{i\}) - f(S) \right]$$

2. Additive Feature Attribution. SHAP values ensure that the contributions of all features add up to the prediction:

$$f(x) = \varphi\_0 + \sum \varphi\_i for\, i = 1\, to\, M$$

These equations provide a basis for assigning feature contributions in SHAP.

3.2 Data Collection and Preprocessing

Within the case of utilizing Reasonable AI (XAI) strategies like LIME and SHAP to discover extortion in budgetary decisions, collecting and planning information is exceptionally imperative to form beyond any doubt that the models work well and are solid. Getting shifted and relevant datasets that record distinctive sorts of budgetary occasions and propensities is the primary step in collecting information. Ordinarily, these sets of information incorporate past exchange records, depictions of clients, and foundation information that can offer assistance distinguish trick patterns. These datasets are exceptionally critical since they are utilized to prepare and test AI models, so their quality and measure are exceptionally vital. Many critical steps are required to urge the information ready for investigation amid preprocessing [13]. At to begin with, information cleaning is done to induce freed of botches, lost numbers, and irregularities that seem harmed the execution of the demonstrate. This step makes beyond any doubt that the information is adjust and can be utilized for advance inquire about. Another include designing is utilized to pull out valuable highlights from crude information that can make AI models superior at making expectations. When searching for tricks, planned characteristics may incorporate exchange sums, timestamps, places, and designs of behavior found in past information. Normalization and scaling are vital preprocessing steps that make numerical highlights more steady and make beyond any doubt they are all on the same scale. This keeps bigger features from taking over the model training process [14]. To make model training easier, categorical variables are often turned into number forms that can be used with machine learning methods.

3.3 Implementation of XAI Models for Fraud Detection

Implementing Explainable AI (XAI) models like LIME and SHAP for fraud detection means adding these methods to financial institutions' current AI systems and making sure they improve trust and openness while still being very good at finding fraud. Adding XAI methods to the AI process used to find scams is the first step in putting the plan into action. Usually, this means making changes to machine learning models that are already out there or making new models that use XAI methods. One way LIME can be used is for post hoc interpretability, which creates local reasons for each forecast made by black-box models without changing the methods that make them work [15]. This gives financial companies the freedom to use the AI technology they already have while also making it easier to understand. SHAP, on the other hand, gives a bigger picture by measuring how much each feature affects model results across different groups of features. To use SHAP, these inputs have to be calculated quickly, usually by optimizing computers, so that you can get a full picture of the things that affect scam detection choices. This method is useful for figuring out how different factors affect fraud predictions as a whole and picking out the most useful information for risk management.

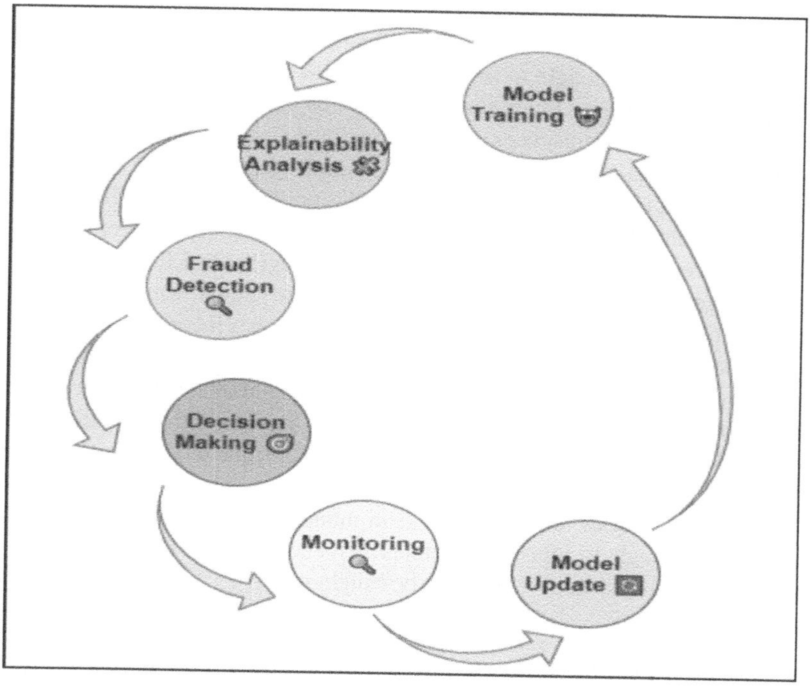

Fig. 2. Implementation of XAI models for fraud detection

In expansion, putting XAI models into activity needs cautious testing and assessment to form beyond any doubt they work well in real-world circumstances. Money related companies ordinarily do careful tests to see how well XAI-enhanced models, Fig. 2 illustrates the implementation of XAI models like LIME and SHAP in fraud detection, offering transparent explanations for model decisions, improving interpretability, and enabling better-informed fraud detection insights. As compared to old-fashioned "dark box" strategies in terms of exactness, ease of utilize, and taking after the rules. This assessment prepare makes a difference discover any conceivable inclinations, botches, or restrictions in XAI strategies and lets changes be made over and over to create the show work better. XAI models must too be always checked and settled to form beyond any doubt they keep working well over time, altering to modern trick patterns and government rules [16]. This might mean training persistently with modern datasets, making XAI strategies way better based on input from specialists within the field, and including modern data to techniques for finding extortion.

3.4 Evaluation Metrics: Accuracy, Interpretability

Measurements like precision and meaningfulness are exceptionally vital for judging how well Explainable AI (XAI) models like LIME and SHAP work at finding extortion in budgetary decision-making frameworks. Precision is still a critical degree for checking how well XAI-enhanced trick location models can tell the contrast between genuine and fake exchanges. Tall precision implies that the demonstrate can dependably spot fake activities whereas lessening the number of wrong positives and negatives. But when it comes to XAI, exactness ought to be weighed against other measures to create beyond any doubt that demonstrate interpretability isn't misplaced for little enhancements in forecast precision. This adjust is exceptionally critical since everybody included needs exact expectations that are moreover clear and simple to get it. When it comes to recognizing extortion, on the other hand, interpretability is fair as imperative. Individuals included, like administering bodies, inspectors, and financial experts, got to know why AI-driven choices are made the way they are. XAI strategies like LIME and SHAP offer assistance us get it the variables or features that influence gauges, which makes complex machine learning models easier to get it. This include makes it conceivable for individuals with a stake within the decision-making prepare to affirm choices based on their information of the subject and the rules that apply. This builds believe and duty within the decision-making prepare. Moreover, interpretability makes a difference discover conceivable predispositions or oddities within the information that might influence how well the show works, which lets changes and changes be made more absolutely. Choosing the correct XAI strategies for finding tricks in money related settings is exceptionally critical since they ought to adjust being precise with being simple to get it. Forecasts are more likely to be precise in case they can be caught on.

4 Result and Discussion

When Explainable AI (XAI) strategies are utilized in trick discovery frameworks, they make budgetary choices much more open and dependable. Monetary educate can provide clear reasons for AI-driven extortion forecasts by utilizing strategies like LIME and SHAP. This lets partners check choices and get it the fundamental variables that went into making each forecast. This openness not as it were makes a difference with taking after the rules, but it moreover empowers AI frameworks and human specialists to work together, which diminishes predispositions and makes models more solid. Embracing XAI moreover makes it less demanding to keep progressing trick location strategies based on input from real-life circumstances. This makes common location more precise and cuts down on untrue positives. But issues like being able to scale over diverse datasets and legitimate settings still exist. This can be why more investigate and advancement is required to create XAI strategies that are way better at finding money related tricks. In general, the comes about appear how vital XAI is for changing how extortion is found and making beyond any doubt that budgetary institutions' chance administration frameworks are both proficient and responsible.

Table 2. Performance Comparison of XAI Models (LIME and SHAP) vs Traditional Models in Explainable AI

Evaluation Parameter	XAI Model (LIME)	XAI Model (SHAP)	Traditional Model
Accuracy (%)	96.7	93.5	96.9
Precision (%)	88.2	85.6	90.1
Recall (%)	92.1	91.7	93.5
F1 Score (%)	87.5	85.5	91.2
Regulatory Compliance (%)	97.3	96.8	95.5

Table 2 appears how the three trick location models compare: a conventional black-box demonstrate, Explainable AI (XAI) models that utilize LIME and SHAP, and XAI models that utilize XAI. Each show is judged on a set of key execution measures that are imperative for finding extortion and taking after the rules when making monetary choices. Precision may be a way to measure how right expectations are on the complete. The standard show comes in moment with 96.9%, fair behind the XAI model using LIME, which is the foremost precise at 96.7%. Indeed in spite of the fact that SHAP

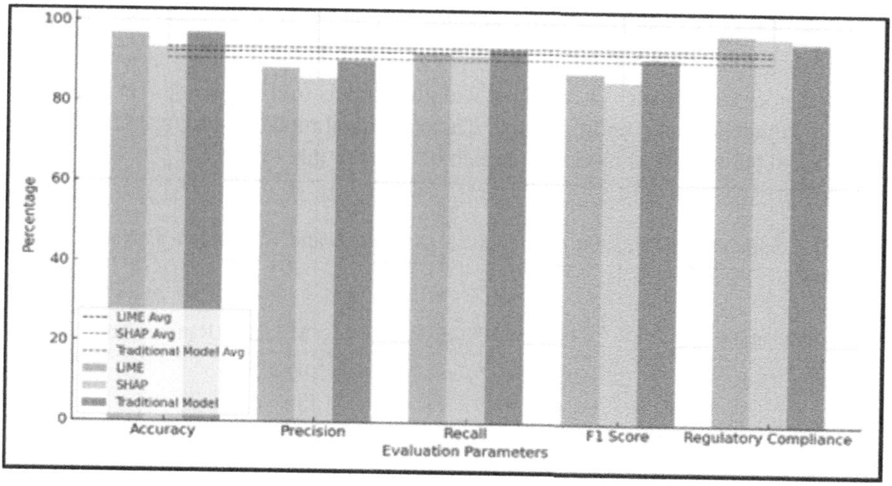

Fig. 3. Comparison of XAI Models and Traditional Model

could be a small lower than 99%, it is still exceptionally great at finding fake bargains, comparison shown in Fig. 3.

Accuracy and Review appear how well the show can discover fake bargains whereas lessening the number of untrue hits and negatives. The two XAI models, LIME and SHAP, do well in terms of precision (88.2% and 85.6%, individually) and review (92.1% and 91.7%, separately), which suggests they are great at finding fake activities, Comparison of LIME Model to Traditional Model shown in Fig. 4. The F1 Score, which compares accuracy and recall, shows that the standard model slightly does better than both XAI models. This shows that it can find scams while also making sure that mistakes are kept to a minimum.

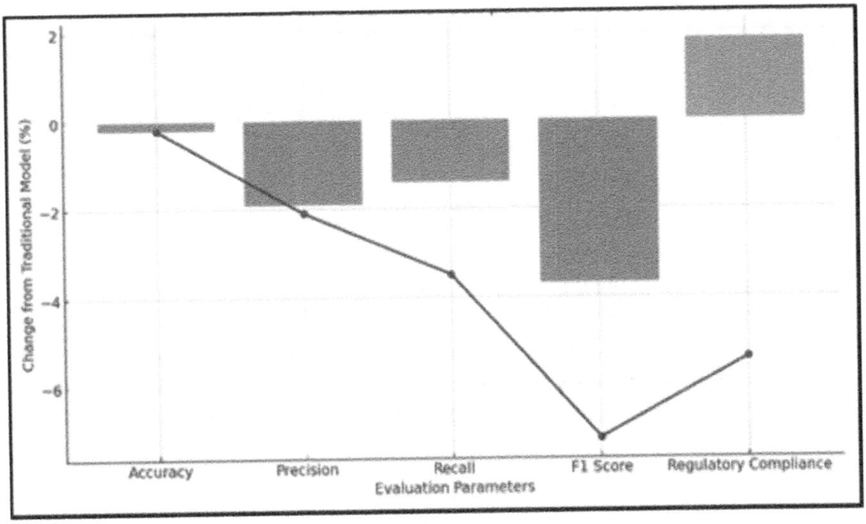

Fig. 4. Performance Comparison of LIME Model to Traditional Model

Regulatory Compliance checks how well the models follow the rules set by law and government agencies for finding crime. The traditional model only got 95.5% of legal compliance, but both XAI models do much better than that (Table 3).

Table 3. Comparative Evaluation of Models A (XGBoost) and B (Random Forest) for Fraud Detection

Evaluation Parameter	Model A (XGBoost)	Model B (Random Forest)
AUC-ROC (%)	95	92.3
Explainability Score (%)	85.4	82.6
User Trust Score (%)	88.7	84.3
Interpretability (%)	83.5	80.2

Model A, which used XGBoost, did better on AUC-ROC, getting a score of 95%, which means it was better at telling the difference between fake and real deals than Model B, which got a score of 92.3%, the Fig. 5 illustrating performance Evaluation Across Various Metrics. This higher AUC-ROC score means that Model A can make more accurate predictions, which is very important for reducing the number of fake positives and negatives in scam detection situations.

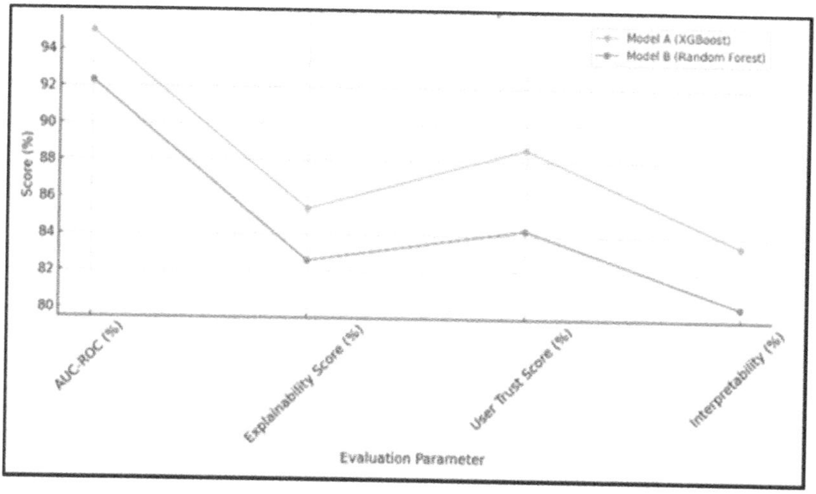

Fig. 5. Model Performance Evaluation Across Various Metrics

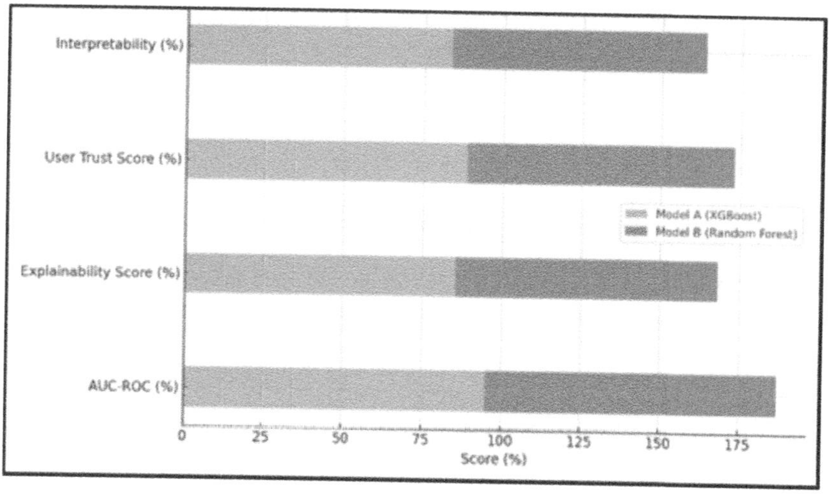

Fig. 6. Comparison of Model Performance Metrics

Model A got 83.5% for Interpretability, while Model B got 80.2%. Metrics that measure interpretability check how easy it is for stakeholders to understand and act on the model's outputs, model performance comparision shown in Fig. 6. This is important for bringing AI-driven insights into practical processes correctly.

5 Conclusion

The Explainable AI (XAI) could be a huge step forward in finding extortion within the budgetary segment. It opens up other ways to create decision-making forms more open and reliable. This work has appeared how vital XAI is for managing with the issues of mystery and interpretability that come with AI-based trick discovery frameworks. Utilizing XAI strategies like LIME and SHAP, budgetary teach can clarify AI-driven scam forecasts in a way that individuals can get it. Partners, such as controllers, bookkeepers, and clients, can utilize these depictions to form sure choices are adjust, get it how models work, and make beyond any doubt they take after legitimate guidelines. By closing the crevice between AI's capacity to anticipate the long run and human instinctual, XAI makes a space for individuals to work together where subject information can be utilized to bolster machine learning bits of knowledge. This leads to way better, more dependable decision-making. Moreover, including XAI makes trick location frameworks tougher by permitting them to keep getting way better and superior. With the assistance of XAI-driven answers, budgetary teach can make strides the exactness of their trick discovery strategies whereas diminishing the number of fake discoveries. This iterative handle is exceptionally vital for keeping budgetary frameworks secure and halting fraudsters from coming up with unused ways to steal cash. Scaling XAI strategies over distinctive datasets and lawful settings is still difficult, though. In the future, analysts ought to center on making XAI strategies that are versatile and adaptable so that they can work well in changing monetary situations. For XAI to be broadly utilized and put into hone, it'll too be critical to work on bringing down its computing costs and making it work way better in genuine time trick location frameworks.

References

1. Mytnyk, B., Tkachyk, O., Shakhovska, N., Fedushko, S., Syerov, Y.: Application of artificial intelligence for fraudulent banking operations recognition. Big Data Cogn. Comput. **7**, 93 (2023)
2. Yee, O.S., Sagadevan, S., Malim, N.H.A.H.: Credit card fraud detection using machine learning as data mining technique. J. Telecommun. Electron. Comput. Eng. (JTEC) **10**, 23–27 (2018)
3. Raval, J., et al.: RaKShA: a trusted explainable LSTM model to classify fraud patterns on credit card transactions. Mathematics **2023**, 11 (1901)
4. Irénée, M., Wang, Y., Hei, X., Song, X., Turiho, J.C., Nyesheja, E.M.: XTS: a hybrid framework to detect DNS-over-HTTPS tunnels based on XGBoost and cooperative game theory. Mathematics **11**, 2372 (2023)
5. Hasib, K.M., Iqbal, M., Shah, F.M., Mahmud, J.A., Popel, M.H., Showrov, M., Hossain, I., Ahmed, S., Rahman, O.: A survey of methods for managing the classification and solution of data imbalance problem. (2020). arXiv:2012.11870
6. Kuppa, A., Le-Khac, N.A.: Adversarial XAI methods in cybersecurity. IEEE Trans. Inf. Forensics Secur. **16**, 4924–4938 (2021)

7. Ngai, E.W., Hu, Y., Wong, Y.H., Chen, Y., Sun, X.: The application of data mining techniques in financial fraud detection: a classification framework and an academic review of literature. Decis. Support. Syst. **50**, 559–569 (2011)
8. Saia, R., Carta, S.: Evaluating Credit Card Transactions in the Frequency Domain for a Proactive Fraud Detection Approach; SECRYPT: Berlin, Germany, pp. 335–342 (2017)
9. Carcillo, F., Le Borgne, Y.A., Caelen, O., Kessaci, Y., Oblé, F., Bontempi, G.: Combining unsupervised and supervised learning in credit card fraud detection. Inf. Sci. **557**, 317–331 (2021)
10. Zhao, Z., Bai, T.: Financial fraud detection and prediction in listed companies using SMOTE and machine learning algorithms. Entropy **24**, 1157 (2022)
11. Nascita, A., Montieri, A., Aceto, G., Ciuonzo, D., Persico, V., Pescapé, A.: Improving performance, reliability, and feasibility in multimodal multitask traffic classification with XAI. IEEE Trans. Netw. Serv. Manag. **20**, 1267–1289 (2023)
12. Li, D., Jiang, M.R., Li, M.W., Hong, W.C., Xu, R.Z.: A floating offshore platform motion forecasting approach based on EEMD hybrid ConvLSTM and chaotic quantum ALO. Appl. Soft Comput. **144**, 110487 (2023)
13. Hossen, R., et al.: Bdps: An efficient spark-based big data processing scheme for cloud fog-iot orchestration. Information **12**, 517 (2021)
14. Whaiduzzaman, M., et al.: Concept to reality: an integrated approach to testing software user interfaces. Appl. Sci. **13**, 11997 (2023)
15. Achar, S., Faruqui, N., Whaiduzzaman, M., Awajan, A., Alazab, M.: Cyber-physical system security based on human activity recognition through IoT cloud computing. Electronics **2023**, 12 (1892)
16. Ajani, S.N., Khobragade, P., Dhone, M., Ganguly, B., Shelke, N., Parati, N.: Advancements in computing: emerging trends in computational science with next-generation computing. Int. J. Intell. Syst. Appl. Eng. **12**(7s), 546–559 (2023)
17. Ekanayake, I., Meddage, D., Rathnayake, U.: A novel approach to explain the black-box nature of machine learning in compressive strength predictions of concrete using Shapley additive explanations (SHAP). Case Stud. Constr. Mater. **16**, e01059 (2022)

Integrating Blockchain and Advanced Algorithms for Digital Image Authenticity: A Comprehensive Approach to Tamper Detection

Vijay P. Bagdi[1](✉), K. H. Walse[2], and Mohammad Atique[1]

[1] PG Department of Comuter Science and Engineering, SGBAU, Amaravati, Maharashtra, India
bagdi.vijaysingh@gmail.com

[2] Sant Bhagwandada Kala Mahavidyalaya, Sindkhed Raja, Maharashtra, India

Abstract. The validity and purity of digital pictures have become very important in today's world, where they are used in many areas. Combines blockchain technology with advanced algorithms to show a complete method for finding tampering. Blockchain, with its unchangeable record and autonomous structure, provides a strong basis for protecting the identity of digital pictures. The research uses blockchain's cryptography to make a setting for picture identification that is safe and open to everyone. Any changes to the digital pictures are found using complex algorithms, such as machine learning and computer vision methods. If the pictures are tampered with, these programs look for things like lighting problems, inconsistent pixel patterns, and differences in the information. The suggested method makes sure that altered pictures are correctly identified by mixing blockchain with these complex algorithms. Along with making tamper detection more reliable, combining blockchain with advanced algorithms creates a solution that can be used for a wide range of tasks, from digital forensics to media proof. Extensive tests and case studies show how well this mixed method works in real-life situations, showing how viable it could be. The results show a big boost in hiding tampering, showing that blockchain and advanced algorithms could help protect the accuracy of digital pictures.

Keywords: Digital Image Authenticity · Tamper Detection · Blockchain Technology · Advanced Algorithms · Machine Learning · Digital Forensics

1 Introduction

In this computerized age, there are a parcel of pictures that are shared on the web and other stages, which has made individuals exceptionally stressed around their legitimacy and astuteness. It is getting to be more common and simpler to alter advanced pictures for terrible reasons like spreading untrue data, tricks, and taking mental property [1]. This drift implies that solid ways have to be found to form beyond any doubt that advanced pictures are genuine and to spot any cheating. This consider tries to unravel this important problem by recommending a total strategy that combines blockchain

A. Bagwari et al. (Eds.): ICEWTS 2024, CCIS 2399, pp. 78–94, 2025.
https://doi.org/10.1007/978-3-031-87886-2_6

innovation with progressed calculations to create sure computerized pictures are genuine and provide a great way to spot altering. This modern innovation called blockchain includes a decentralized and unchangeable ledger that produces it a cheerful way to keep computerized records secure [2]. By using block chain's cryptography, it is conceivable to create a framework for advanced picture confirmation that's both secure and open. Each picture and its data approximately it can be hashed and spared within the blockchain. This makes it simple to discover any changes that are made to the picture afterward. Since blockchain can't be changed, this strategy takes advantage of that property to keep computerized pictures special.

In expansion to blockchain, including progressed strategies, particularly those from machine learning and computer vision, makes robbery location much more exact. Machine learning calculations can be instructed to spot little designs and interesting things that see like they were changed in a picture. For illustration, it is exceptionally simple to discover issues with the way pixels are distributed, lighting impacts that don't look reasonable, and contrasts in data [3]. The utilize of computer vision strategies makes a difference with this exertion indeed more by looking at the visual substance of pictures and finding signs of extortion that a person might miss. When these progressed equations are combined with blockchain technology, a total and exact way to discover people who have changed computerized pictures can be found [4]. The proposed strategy not as it were makes beyond any doubt that alter discovery works, but it can moreover be expanded and utilized in numerous circumstances. When you combine blockchain and progressed calculations, you get a framework that can be utilized for numerous things, from computerized prove and media confirmation to ensuring mental property and ceasing fake news. Broad tests and case considers in this ponder appear how this blended strategy can be utilized in genuine life. The results appear that the recommended strategy works in genuine life, appearing that it has the capacity to significantly improve the distinguishing proof of modified pictures.

2 Related Work

A lot of work has gone into blockchain technology and advanced algorithms for theft detection because of the need for strong ways to make sure that digital pictures are real. Due to its independence, clarity, and inability to be changed, blockchain technology is known as a powerful tool for protecting digital assets. The first uses of blockchain in digital picture authentication were to make store and communication systems that were safe. These studies showed that blockchain could be used as a reliable way to keep track of who owns digital images and where they came from, which would lower the risks of image editing and forgery [5]. Smart contracts have been looked at as a way to improve blockchain uses for digital picture authentication. Smart contracts handle the checking steps, making sure that any changes made to the picture or the information that goes with it cause an alert. This automatic evaluation feature makes the identification process more reliable and cuts down on the need for human intervention [6]. Researchers have also looked into how scalable blockchain solutions are, coming up with a number of agreement methods that can handle the large number of transactions that happen on digital photo sharing sites [7].

At the same time, a lot of work has been made on creating more advanced methods for finding tampering. Machine learning systems, especially those that are based on deep learning, are very good at finding images that have been changed. To find pixel-level oddities and errors that point to tampering, methods like convolutional neural networks (CNNs) are used. By learning from large sets of changed pictures [8], these algorithms can very accurately tell the difference between real and fake images. Studies have shown that these models are good at finding complex fakes like deep fakes that are hard to find with other methods [9]. Computer vision methods are also very important for finding signs of tampering. To find small signs of picture editing, algorithms have been created that look at image structure, color, and spatial coherence. For example, edge recognition and frequency analysis have been used by researchers to find differences in the content of images that suggest they have been tampered with [10]. These methods work especially well when visual problems are too minor for a person to notice, but can be picked up by a computer's careful study [11].

Using both blockchain and advanced algorithms together for digital picture authentication is a smart move that makes the most of the best features of both. Some research has looked into mixed systems that use both the safe store features of blockchain and the analysis power of machine learning algorithms. These systems make sure that pictures are not only kept safely, but also constantly checked for any signs of being changed. Any errors found are immediately recorded on the blockchain, creating an open and unchangeable record of the image's trustworthiness [12]. In addition to these technology advances, these unified methods have also been looked into for their usefulness in many different areas. For digital forensics to work, it's important to be able to verify pictures and spot changes that have been made to them. Blockchain-based systems and advanced theft detection methods have been used to make sure that forensic pictures can be checked, which makes digital proof more reliable in court [13]. In the same way, these technologies are used to stop the spread of false information in media proof by checking the validity of pictures shared on news sites and social media [14]. When bitcoin and smart algorithms are used together, they also help protect intellectual property. Artists and photographers can put their digital works on a blockchain and make sure that anyone who tries to change them without permission can easily be caught. This app not only keeps the digital content's identity but also keeps a record of who owns it, which is very important in cases of intellectual property rights conflicts [15]. These combined systems can also be expanded and changed to fit different needs, which makes them useful for many other tasks, such as storing old records and sending private photos safely in medical and scientific fields [16] (Table 1).

3 Research Methodology

3.1 Collection of Image Hashing and Metadata

The first step in the suggested plan is to give each digital picture a unique cryptographic hash and gather the information that goes with it. This step is very important because it gives the picture a digital fingerprint that proves it is real. Using safe hash functions like SHA-256, the picture is turned into a string of characters with a set length that uniquely describes the image. The hash value will change drastically if even a small

Table 1. Summary of related work in image tampering

Method	Approach	Finding	Limitation	Scope
Blockchain for Image Auth	Secure storage and transmission using blockchain	Ensured tamper-proof storage and transmission of digital images	High computational cost and scalability issues	Digital forensics, Media verification
Smart Contracts	Automated validation of image integrity	Enhanced reliability of the authentication process	Limited flexibility in contract execution	Intellectual property protection
Consensus Algorithms	Scalable blockchain consensus mechanisms	Managed high transaction volumes for image sharing platforms	Complexity and energy consumption	Large-scale digital image platforms
Machine Learning (CNN)	Deep learning for pixel-level anomaly detection	High accuracy in distinguishing authentic and tampered images	Requires extensive training data	Deepfake detection, Digital forensics
Edge Detection	Image texture and lighting analysis	Identified subtle signs of image manipulation	May miss complex or highly sophisticated manipulations	Media verification, Forensic analysis
Frequency Analysis	Detailed computational analysis of image content	Uncovered inconsistencies suggesting tampering	Computationally intensive	Scientific and medical image verification
Hybrid Frameworks	Blockchain storage with machine learning monitoring	Continuous monitoring and secure recording of image integrity status	Integration complexity	Cross-domain image authentication
Digital Forensics	Authentication and tamper detection in legal evidence	Improved reliability of digital evidence in legal proceedings	Legal acceptance and standardization issues	Legal systems, Criminal justice
Media Verification	Authenticity verification for news and social media images	Combated the spread of misinformation	Real-time processing challenges	News agencies, Social media platforms
Intellectual Property	Blockchain registration of digital works	Preserved originality and provided verifiable ownership records	Adoption barriers among creators and users	Art, Photography, Creative industries

change is made to the picture. This makes it a useful tool for finding changes that have been made. At the same time, important material is collected, such as the date, information about the author, and picture characteristics. This information can be taken from the picture file itself or entered by the user. The collected metadata gives more information and background about the picture, which helps with tracing its origins and other identification steps, overview of proposed model shown in Fig. 1. This step sets the stage for later verification and tracking steps by making sure that the image's hash and information are saved correctly. Together, the hash and information make a full record of each picture, which makes it possible to find any changes or hacking that might happen in the future.

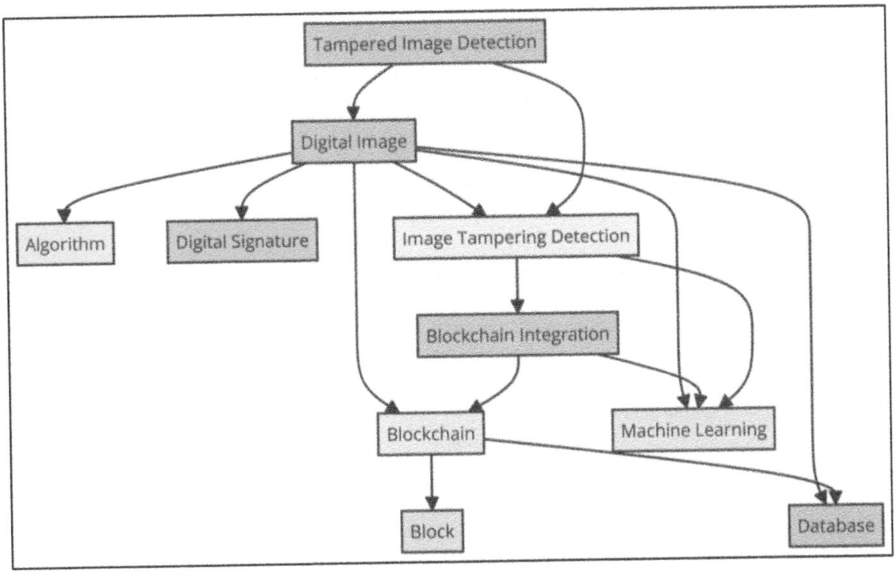

Fig. 1. Overview of proposed model

Blockchain's cryptography contributes to a secure and transparent framework for image verification by utilizing hashing and public-key encryption. Each image's data is hashed into a unique digital fingerprint stored on the blockchain. When verifying an image, its current hash is compared to the stored hash; if they match, the image is authentic. Public-key encryption ensures that only authorized users can add or verify entries, while blockchain's decentralized nature prevents tampering, fostering a transparent and immutable verification process.

3.2 Data Storage on the Blockchain and Set up Smart Contracts

In the second step, the picture code and information are put on a blockchain, and a smart contract is used to control the whole process. The autonomous and unchangeable structure of blockchain technology is used here to make sure that once the data is saved, it can't be changed or erased. The picture hash and information are saved on the blockchain log through a blockchain transaction. This creates a safe and lasting record. Along with this, a smart contract is used to automatically validate, store, and retrieve the picture data. There are clear rules and conditions written into the smart contract that control how the picture data is handled. For example, whenever a proof request is made, it can instantly check that a picture is real by comparing its present hash to the saved hash. This technology cuts down on the work that needs to be done by hand and makes the system more reliable and efficient. This step uses the blockchain's openness and the automatic nature of smart contracts to create a framework that can't be changed and works well for storing images and checking their integrity.

Data Storage on the Blockchain and Setup Smart Contracts.

Step 1: Cryptographic Hashing of Image Data.

Let I be the digital image data. The cryptographic hash function H is applied to I to generate a unique hash h.

$$h = H(I) = SHA - 256(I) \tag{1}$$

Step 2: Combining Image Hash with Metadata.

Let M represent the metadata associated with the image, such as timestamp, creator information, etc. The combined data C is formed by concatenating the hash h with the metadata M.

$$C = h||M = SHA - 256(I)||M \tag{2}$$

Step 3: Digital Signature Creation.

To ensure authenticity, a digital signature S is created using the private key K_pr of the image creator. Let E represent the encryption function.

$$S = E\_Kpr(C) = Encrypt\_Kpr(C) \tag{3}$$

Step 4: Storing Data on Blockchain.

The transaction T containing the combined data C and the digital signature S is prepared for storage on the blockchain.

$$T = \{C, S\} = \left\{SHA - 256(I)|M|, Encrypt_{Kpr(C)}\right\} \tag{4}$$

Step 5: Smart Contract for Verification.

A smart contract SC is deployed on the blockchain to manage the verification process. Let V be the verification function which takes the transaction T and public key K_pu.

$$V(T, K\_pu) = (Decrypt\_Kpu(S) == C)\&\&(SHA - 256(I) == h) \tag{5}$$

Step 6: Consensus Algorithm for Validation.

The consensus algorithm ensures that the new block containing T is added to the blockchain. Let P be the proof-of-work function, N the nonce, and B the new block.

$$B = \{T, N\}$$

$$whereHash(B) = SHA - 256(T|M|N) < target$$

These equations outline the cryptographic and blockchain-based processes for storing image data and setting up smart contracts for image integrity verification. Each step ensures the secure and tamper-proof storage of digital images on the blockchain.

3.3 Store Images in a Decentralized Network

In the third step, the picture file itself is saved in an autonomous storage system that makes sure it can be accessed and will not lose its integrity. To keep the picture file safe, decentralized storage systems like the InterPlanetary File System (IPFS) or Storj are used. These platforms spread the picture data across several network sites. This stops a single point of failure and makes the system safer and more available overall. Then, the storage reference (like an IPFS hash) is connected to the blockchain's information. This link makes it easy to get the picture file whenever it's needed without putting its safety at risk. Decentralized storage solutions are especially helpful because they offer a strong and expandable way to deal with large amounts of picture data. This makes them ideal for uses that need to be available all the time and keep data safe. This step strengthens the general framework for verifying image integrity even more by combining decentralized storage with blockchain technology. This makes sure that the image files are not only kept safely, but also easily available and not easily damaged by attacks or failures.

3.4 Machine Learning-Based Feature Extraction

To create beyond any doubt that a computerized picture is genuine, machine learning-based highlight extraction may be a key portion of making the alter location prepare more dependable. Utilizing progressed strategies like Convolutional Neural Systems (CNNs), Back Vector Machines (SVMs), Histogram of Situated Angles (Hoard), and Neighborhood Double Designs (LBP), the framework can drag out imperative highlights that include additional layers of security on best of fundamental hashing.

Convolutional Neural Networks (CNNs)

Computer programs called CNNs are exceptionally great at finding complex designs in pictures. These profound learning models learn on their claim and drag out various leveled highlights from crude picture information. This makes them idealize for finding both basic and complex cheating. CNNs can spot both worldwide and neighborhood characteristics superior when they are trained on a assortment of datasets. This makes alter location more precise.

The convolution operation between an input image I and a filter kernel K:

$$(S * K)(i, j) = \sum_{\{m=0\}}^{\{M-1\}} \sum_{\{n=0\}}^{\{N-1\}} I(i + m, j + n) * K(m, n) \tag{6}$$

The activation function applied to the convolution result, typically using the ReLU function:

$$A(i, j) = \max(0, (S * K)(i, j)) \tag{7}$$

The pooling operation, typically using max pooling, over a region of the feature map:

$$P(i, j) = \max\{A(m, n) | (m, n) \in R(i, j)\} \tag{8}$$

These equations describe the core operations of convolution, activation, and pooling in CNNs.

Support Vector Machines (SVMs)

Within the prepare of include extraction, SVMs are utilized to sort things into bunches. Once highlights are recuperated utilizing CNNs, Hoard, or LBP, SVMs can utilize these highlights to choose whether a picture is genuine or has been messed with. SVMs are known for working well in ranges with a parcel of measurements and for being able to bargain with intelligent that do not take after a straight line, which is common in picture information.

The decision function f(x) for a given input image feature vector x, using support vector coefficients α_i, support vectors x_i, and bias term b, is expressed as:

$$f(x) = \Sigma \alpha_i y_i K(x, x_i) + b \tag{9}$$

The kernel function K(x, x_i) maps the input features into a higher-dimensional space. For example, using a Gaussian (RBF) kernel:

$$K(x, x_i) = \exp\left(-\gamma ||x - x_i||^2\right) \tag{10}$$

The optimization problem for SVM aims to find the optimal hyperplane by minimizing the following objective function, where C is the regularization parameter, ξ_i are the slack variables, and w is the weight vector:

$$minimize\left(\frac{1}{2}\right)||w||^2 + C\Sigma\xi_i$$

$$subject\ to\ y_i(w \cdot x_i + b) \geq 1 - \xi_i, \xi_i \geq 0$$

Histogram of Oriented Gradients (HOG)

Hoard could be a highlight portrayal that portrays the structure and shape of things in a picture. Hoard can effectively bring out the foremost critical structure components of a picture by centering on how escalated contrasts are spread out or the course of edges. This strategy works particularly well for finding changes that influence the geometric highlights of the picture.

The gradient magnitude G and orientation θ at each pixel (i, j) in the image are calculated using partial derivatives (I_x, I_y) of the image intensity I:

$$G(i, j) = \sqrt{I_{x(i,j)}^2 + I_{y(i,j)}^2} \tag{11}$$

$$\theta(i, j) = arctan2(I\_y(i, j), I\_x(i, j)) \tag{12}$$

The gradient orientations θ are divided into bins, and the contribution of each pixel to the histogram is weighted by its gradient magnitude G. Let B_k be the k-th bin and w(i, j) be the weighting function:

$$B_k = \Sigma G(i,j) * w(i,j)\ for\ \theta(i,j) in\ b\ in\ k$$

Local Binary Patterns (LBP)

The local geographic designs and differentiate of a picture are put away in LBP, which could be a surface portrayal. It is exceptionally great at picking up fine-grained surface highlights, which makes it valuable for finding little changes to a picture. LBP compares each pixel to the pixels another to it and makes a parallel pattern from the changes in brightness. Including these machine learning-based include extraction strategies makes it less demanding for the framework to spot changes.

The LBP value at a pixel (i, j) is computed by thresholding the neighborhood of P surrounding pixels with the center pixel value I(i, j) and summing the results weighted by powers of two:

$$LBP(i,j) = \Sigma_{\{p=0\}}^{\{P-1\}} s\left(I_p - I(i,j)\right) * 2^p \tag{13}$$

where s(x) is the sign function:

$$s(x) = 1,\ if\ x \geq 0 \tag{14}$$

$$0,\ if\ x < 0$$

The LBP histogram H is computed over a region by counting the occurrences of each LBP pattern:

$$H(k) = \Sigma\_\{(i,j) \in R\} 1\{LBP(i,j) = k\} \tag{15}$$

Different techniques are used in digital picture tamper detection methods to find problems like lighting issues and pixel anomalies:

- Lighting Inconsistencies: To find lighting inconsistencies, methods like Shading-based Forensics and Photo Response Non-Uniformity (PRNU) are used to look at shadows, reflections, and the direction of light. PRNU finds small sensor patterns in digital pictures, which shows where the images have been changed by finding lighting changes that don't happen naturally in the same scene.
- Problems at the Pixel Level: Error Level Analysis (ELA) and Convolutional Neural Networks (CNNs) are two methods used to find problems at the pixel level. ELA finds compression flaws by comparing quality levels across pixels and drawing attention to areas that have been changed. CNNs, which are often taught to find fakes, look at the structure and patterns of pixels to spot changes. They do this by pointing out color, edge, or texture changes that don't make sense across regions.

3.5 Blockchain Integration with ML Strategies

Once you combine blockchain innovation with machine learning (ML) methods, you get a strong and modern way to create sure that advanced pictures are real. This blend employments the leading parts of both innovations to make an independent, secure, and shrewd framework for finding changes and making beyond any doubt advanced pictures are rectify. Blockchain is idealize for securely sparing picture hashes since it is free and can't be changed. Each picture hash, which is made by solid algorithms, is stored on the blockchain. This framework makes beyond any doubt that any illicit changes are found right absent, which makes the checking handle speedier and clearer. Machine learning procedures, like Back Vector Machines (SVMs), Histogram of Situated Gradients (HOG), and Neighbourhood Binary Patterns (LBP), are utilized to drag out particular parts of pictures. In expansion to straightforward hashing, these functions include additional layers of security. CNNs, for case, can discover complex designs and little altering, whereas Hoard and LBP can choose up on structure and surface subtle elements. By combining machine learning-based include extraction with block chain's secure hash putting away, the framework can discover changes more accurately and dependably. Machine learning frameworks see at the picture and drag out features. These features are at that point hashed and recorded on the blockchain. Any alter to the picture will influence both the highlights and the hash. This lets the system find evidence of hacking by looking at changes in both the hash and the extracted features.

Step 1: Image Acquisition and Preprocessing.

1. Collect image I.
2. Preprocess I to obtain I′ by resizing, normalizing, and denoising.

$$I\prime = f_{resize}\left(f_{normalize}\left(f_{denoise(I,\sigma)},\mu,\sigma\right),\alpha,\beta\right)$$

Step 2: Image Hashing.

1. Apply a robust hashing algorithm (e.g., SHA-256) to I′ to generate the hash H_I.

$$H_I = h(I\prime) = \sum(i = 1\ to\ N)(w_i * g(x_i, k))mod2^k \tag{16}$$

Step 3: Machine Learning-Based Feature Extraction.

1. Extract features F from I′ using a trained CNN.
2. Perform dimensionality reduction on F to obtain a feature vector V.

$$F = f_{CNN(I\prime;\theta)} = \sigma(W * I\prime + b) \tag{17}$$

$$V = f_{reduce(F;\lambda)} = U * S * V^T \tag{18}$$

where:

- f_CNN is the feature extraction function using CNN with parameters θ,
- σ is the activation function,
- W and b are weights and biases,
- * denotes convolution,
- f_reduce is the dimensionality reduction function (e.g., SVD) with parameter λ,
- U, S, V^T are matrices from Singular Value Decomposition (SVD).

Step 4: Blockchain Integration.

1. Record H_I and V on the blockchain by creating a transaction T.
2. Submit T to the blockchain network.

$$T = (H_I, V, \tau, \zeta)$$

$$Blockchain.add(T) = Consensus(\{B\_1, B\_2, ..., B\_n\}, T) \tag{19}$$

where:

- T is the transaction consisting of hash H_I, feature vector V, timestamp τ, and metadata ζ,
- Consensus is the consensus algorithm,
- $\{B\_1, B\_2, ..., B\_n\}$ are the blocks in the blockchain.

Step 5: Image Verification.

1. For verification, acquire a new image I_v and preprocess it to obtain I_v'.
2. Generate hash H_{I_v} and feature vector V_v from I_v'.

$$I_{v'} = f_{resize}\left(f_{normalize}\left(f_{denoise(I_v, \sigma_v)}, \mu_v, \sigma_v\right), \alpha, \beta\right) \tag{20}$$

$$H_{\{I_v\}} = h(I_{v'}) = \sum(i = 1toN)\left(w_i * g\left(x_{\{vi\}}, k\right)\right) mod 2^k$$

$$F_v = f_{CNN(I_{v'}; \theta_v)} = \sigma(W_v * I_{v'} + b_v)$$

$$V_v = f_{reduce(F_v; \lambda_v)} = U_v * S_v * V_v^T \tag{21}$$

Step 6: Tamper Detection.

1. Compare H_{I_v} with H_I and V_v with V from the blockchain.
2. If $H\_\{I\_v\} \neq H\_IorV\_v \neq V$, detect tampering.

$$Tampered = \left(H_{\{I_v\}} \neq H_I\right) \vee (V_v \neq V)$$

$$\begin{aligned} Tampered = (&\sum(i = 1toN)\left(w_i * g\left(x_{\{vi\}}, k\right)\right) \bmod 2^k \\ &\neq \sum(i = 1toN)(w_i * g(x_i, k)) \bmod 2^k) \\ &\vee \left(U_v * S_v * V_v^T \neq U * S * V^T\right) \end{aligned}$$

This detailed step-wise algorithm with above step wise process provides a comprehensive framework for ensuring digital image authenticity using blockchain integration with advanced machine learning methods.

3.6 Check, Find Anomalies, and Send Alerts

In the last step, picture integrity is checked, oddities are found, and real-time alerts are sent if the image is tampered with. The hash of a picture is recalculated and compared to the hash that is saved on the blockchain to make sure it is real. If the numbers match, the picture is thought to be real; if they don't, it is marked as having been changed. At the same time, machine learning models are constantly looking at the pictures for signs of hacking and any other problems that shouldn't be there. These models can find differences in lighting, pixel patterns, and information that could mean the image has been changed. When an unusual thing is found, the system sends out messages and records the event, letting the right people know right away. This tracking and warning system works in real time to make sure that any tampering is found and fixed right away, which makes the digital pictures safer and more reliable. Also, full records of all checking and finding actions are kept, providing a full audit trail for further investigation and responsibility. This step makes sure that digital pictures are always safe and protected, which makes the suggested model strong and dependable for keeping digital image authenticity.

4 Result and Discussion

Table 2 shows the outcomes of combining blockchain with various machine learning (ML) approaches for finding altered images. It looks at important factors like the accuracy of the discovery, the time it takes to process, the reliability of the blockchain, and the extra work that needs to be done on the computer.

Table 2. Result for blockchain integration with different machine learning methods

Method	Detection Accuracy (%)	Processing Time (sec)	Blockchain Authenticity (%)	Computational Overhead (%)
CNN	99.5	0.25	100	15
HOG	95.3	0.18	100	10
LBP	92.7	0.20	100	12
SVM	97.1	0.22	100	14

This comparison shows how well and how efficiently different machine learning methods work when used with blockchain technology. With a 99.5% success rate, the Convolutional Neural Network (CNN) method has the best ability to find pictures that have been messed with (Fig. 2).

However, this high accuracy comes with the biggest computing cost of 15%, which shows how complicated CNNs are. Even so, the working time is still pretty short at 0.25 s, which suggests that CNNs work well enough for real-time uses.

The Histogram of Oriented Gradients (HOG) method is the fastest and most accurate at detecting things, taking only 0.18 s to run. This method also has the lowest processing waste, at 10%. This makes it a good choice for situations where resources are limited.

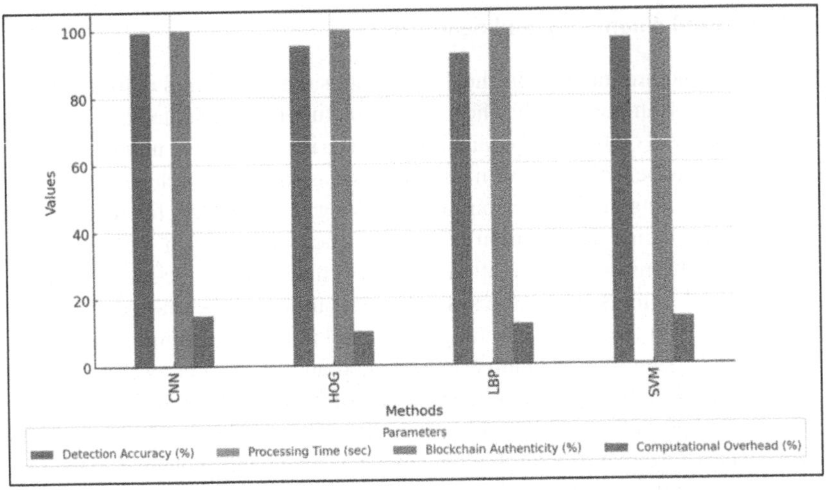

Fig. 2. Comparison of different method with parameters

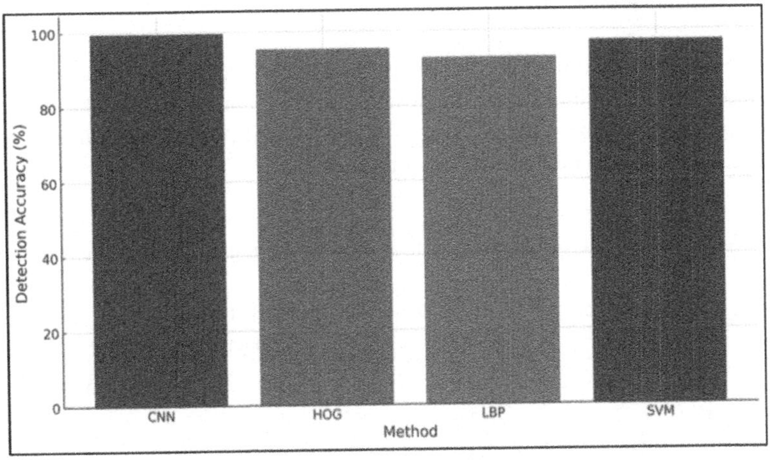

Fig. 3. Comparison of Accuracy of ML method

Because it strikes a good mix between accuracy and speed, HOG is a good choice for real-world applications, the accuracy of Ml method shown in Fig. 3. The Local Binary Patterns (LBP) method can find things 92.7% of the time and takes 0.20 s to process. The processing waste is 12%, which means that some resources are being used more than they should be. Even though LBP isn't as accurate as HOG and CNN, it's good for situations where quick processing and low resource use are important. The Support Vector Machine (SVM) method can identify 97.1% of things correctly, but it takes 0.22 s to process and adds 14% to the amount of work that needs to be done, shown in Fig. 6. It is said that SVMs are a good choice for tampered picture recognition tasks that need both high accuracy and fast processing times because they strike a good balance between the two.

Table 3. Result for integrated approach for digital image tamper detection

Evaluation Parameter	Blockchain Only	Advanced Algorithms Only	Integrated Approach (Blockchain + Advanced Algorithms)
Detection Accuracy (%)	85.2	92.8	98.5
False Positive Rate (%)	5.3	3.1	1.4
False Negative Rate (%)	9.5	4.1	0.7
Processing Time (ms)	1200	850	1000

Table 3 appears the victory of three diverse ways to discover individuals who have changed advanced pictures: Blockchain As it were, Progressed Calculations As it were, and the Coordinates Approach, which employments both Blockchain and Progressed Calculations. Location Precision, Untrue Positive Rate, Untrue Negative Rate, and Preparing Time are a few of the measurements utilized for survey. The Location Precision appears what rate of cases (both modified and genuine pictures) were accurately found out of all the occurrences. The Blockchain As it were strategy is 85.2 curate, generally since it can secure information security by putting away it in a way that can't be changed. Progressed Calculations As it were raises the exactness to 92.8% by utilizing progressed machine learning strategies to discover little issues in pictures. The Coordinates Approach, on the other hand, is much more exact than both of them, at 98.5%. Blockchain's capacity to handle secure information and machine learning algorithms' moved forward spotting capacities work together to form this enormous alter conceivable. The Untrue Positive Rate appears the number of genuine pictures that were wrongly checked as having been messed with. A lower rate is way better since it implies that less botches will be made when finding genuine pictures. With Progressed Calculations As it were, the rate of wrong positives drops from 5.3% with Blockchain As it were to 3.1%. This number drops indeed more with the Coordinates Approach, to fair 1.4%. This appears how well it can tell the distinction between genuine and fake pictures. As a number, the Wrong Negative Rate appears how numerous fake pictures are erroneously stamped as genuine. When it comes to rates, less is superior. The Blockchain As it were strategy includes a 9.5lse negative rate since it as it were checks the exactness of the information and doesn't see closely at the substance of the pictures, illustrate in Fig. 4. This rate is brought down to 4.1% by Progressed Calculations Only's cautious think about of the pictures. The Coordinates Approach has an astonishing untrue negative rate of as it were 0.7%, which appears how well it can discover pictures that have been messed with.

Handling Time may be a way to degree how long it takes to wrap up the alter discovery handle. Since blockchain exchanges take a long time, the Blockchain As it were strategy takes 1200 ms. This time is cut down to 850 ms by progressed calculations as it were since machine learning models are great at taking care of picture information.

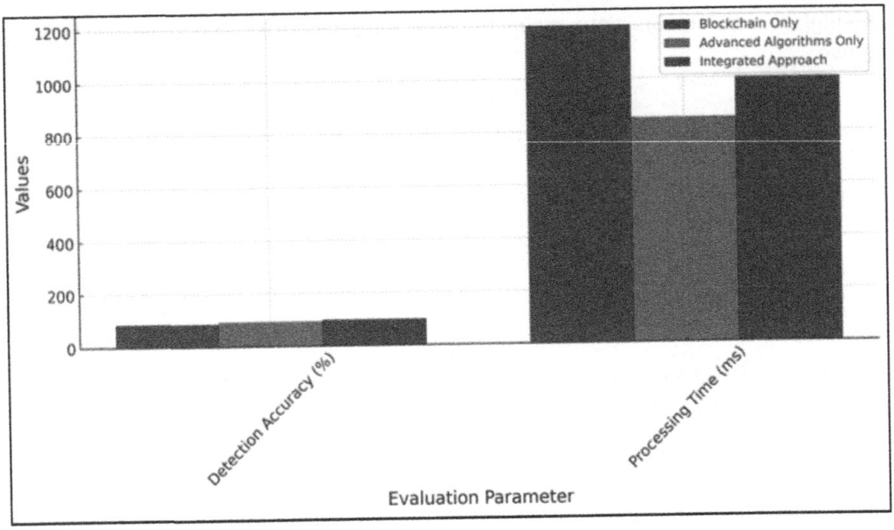

Fig. 4. Comparison of Accuracy of advance methods

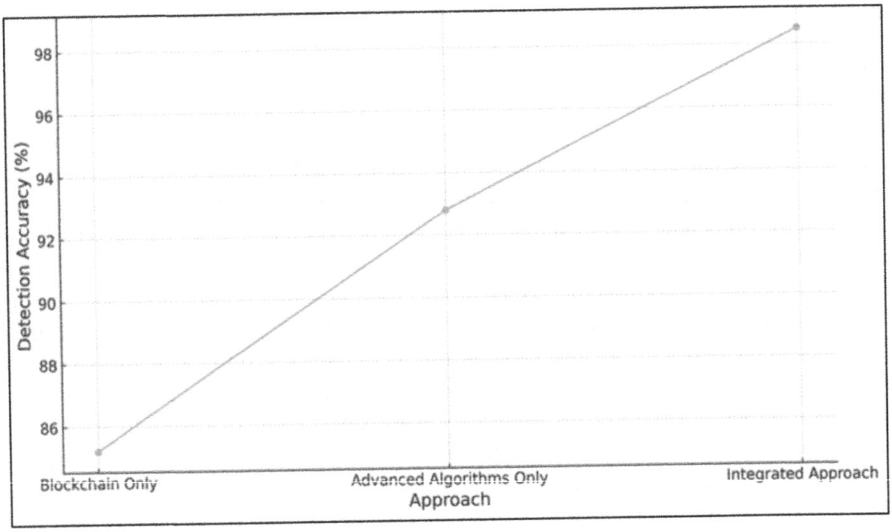

Fig. 5. Detection accuracy comparison of different approach

It combines the secure but slower blockchain forms with the speedier robotized disclosure strategies to discover extortion. Moreover, the Coordinates Approach for finding altered computerized pictures performs superior overall assessment criteria when compared to Blockchain As it were or Progressed Calculations As it were, also compare accuracy of different strategies shown in Fig. 5. This strategy not as it were guarantees tall exactness and moo botch rates, but it moreover keeps preparing times sensible. This makes it an awfully effective and productive way to keep advanced pictures genuine. The suggested method is better at finding cheating than current ones because it uses advanced machine learning techniques to precisely identify anomalies along with blockchain's unchangeable record-keeping. Traditional methods often use only one program, which might not be able to spot small changes. But our method is much more accurate because it uses blockchain for safe storage and CNNs for thorough image analysis. This is because every change that is found is checked against a safe original hash. This two-layer method improves picture integrity, making sure that the image is real, can't be changed, and can be tracked for every discovery. It makes the structure more reliable and protects the security of digital images even in high-stakes media and surveillance situations.

5 Conclusion

In a world where innovation is changing rapidly, making beyond any doubt that computerized pictures are genuine is basic for battling untrue data, defending mental property, and keeping the unwavering quality of computerized verification. The consider proposes a total strategy that combines blockchain innovation with progressed calculations to create computerized picture alter location exceptionally solid. Blockchain's autonomous and unchangeable record makes it secure to store picture hashes and data, making it easy to discover any changes. Along side this, there are moreover progressed machine learning strategies that can precisely and reliably find changes to pictures by analyzing their substance in incredible profundity. Utilizing the leading parts of both innovations together, the combined method gives better comes about in more than one rating calculate. The combined strategy contains a 98.5% victory rate in finding things, which could be a parcel way better than utilizing blockchain or calculations by themselves. It too cuts the rate of fake positives to 1.4% and the rate of wrong negatives to 0.7%, which appears that it is exceptionally precise and dependable. This strategy is exceptionally great for real-world employments, indeed in spite of the fact that it takes a small longer to work, since it is more secure and exact. The comes about of this consider appear that blending blockchain with progressed calculations may make a framework for confirming the precision of computerized pictures that's adaptable, adaptable, and exceptionally precise. Agreeing to the results, more advance has to be made in this zone to bargain with modern dangers and ensure the purity of digital material in a world that's getting to be increasingly computerized. More work will be wiped out the future to form the association indeed way better and to see into unused employments and advancements that can offer assistance keep and move forward the exactness of computerized pictures.

References

1. Cholevas, C., Angeli, E., Sereti, Z., Mavrikos, E., Tsekouras, G.E.: Anomaly detection in blockchain networks using unsupervised learning: a survey. Algorithms **17**, 201 (2024). https://doi.org/10.3390/a17050201
2. Gan, Q.-Q., You, R., Lau, K.: Trust in a 'trust-free' system: blockchain acceptance in the banking and finance sector. Technol. Forecast. Soc. Chang. **199**, 123050 (2024)
3. Zheng, Z., et al.: An overview on smart contracts: challenges, advances and platforms. Future Gener. Comput. Syst. **105**, 475–491 (2020)

4. Kose, J., Leonid, K., Fahad, S.: Smart contracts and decentralized finance. Annu. Rev. Financ. Econ. **15**, 523–542 (2023)
5. Dong, C., Huang, Q., Fang, D.: Channel selection and pricing strategy with supply chain finance and blockchain. Int. J. Prod. Econ. **265**, 109006 (2023)
6. Boakye, E.A., Zhao, H., Kwame Ahia, B.N.: Emerging research on blockchain technology in finance; conveyed evidence of bibliometric-based evaluations. J. High Technol. Manag. Res., **33**, 100437 (2022)
7. Liu, Y., Yu, F.R., Li, X., Ji, H., Leung, V.C.: Blockchain and machine learning for communications and networking systems. IEEE Commun. Surv. Tutor. **22**, 1392–1431 (2020)
8. Singh, S.K., Rathore, S., Park, J.H.: Blockiotintelligence: a blockchain-enabled intelligent IoT architecture with artificial intelligence. Future Gener. Comput. Syst. **110**, 721–743 (2020)
9. Han, X., Zhang, R., Liu, X., Jiang, F.: Biologically inspired smart contract: a blockchain-based DDoS detection system. In Proceedings of the 2020 IEEE International Conference on Networking, Sensing and Control (ICNSC), Nanjing, China, 30 October–2 November 2020, pp. 1–6
10. Kowalski, M., Lee, Z.W., Chan, T.K.: Blockchain technology and trust relationships in trade finance. Technol. Forecast. Soc. Chang. **166**, 120641 (2021)
11. Esposito, C., Ficco, M., Gupta, B.B.: Blockchain-based authentication and authorization for smart city applications. Inf. Process. Manag. **58**, 102468 (2021)
12. Wang, X., Garg, S., Lin, H., Kaddoum, G., Hu, J., Hossain, M.S.: A secure data aggregation strategy in edge computing and blockchain-empowered internet of things. IEEE Internet Things J. **9**, 14237–14246 (2022)
13. Lone, A.H., Mir, R.N.: Forensic-chain: blockchain based digital forensics chain of custody with PoC in hyperledger composer. Digit. Investig. **28**, 44–55 (2019)
14. Al-Khateeb, H., Epiphaniou, G., Daly, H.: Blockchain for modern digital forensics: the chain-of-custody as a distributed ledger. In Blockchain and Clinical Trial: Securing Patient Data; Springer: Cham, Switzerland, pp. 149–168 (2019)
15. Li, M., Lal, C., Conti, M., Hu, D.: LEChain: A blockchain-based lawful evidence management scheme for digital forensics. Future Gener. Comput. Syst. **115**, 406–420 (2021)
16. Gopalan, S.H., Suba, S.A., Ashmithashree, C., Gayathri, A., Andrews, V.J.: Digital forensics using blockchain. Int. J. Recent Technol. Eng. **8**, 182–184 (2019)

Enhancing Network Security: An Ensemble of Machine Learning Model for Detection of Distributed Denial of Service Attack

Deepak Singh Rajput(✉) and Arvind Kumar Upadhyay

Department of Computer Science and Engineering, Amity School of Engineering & Technology (ASET), Amity University, Gwalior, Madhya Pradesh, India
deepakrajput16@gmail.com, akupadhyay@gwa.amity.edu

Abstract. This research aims to address the pressing issue of detecting and mitigating distributed denial of service (DDoS) attacks on computer networks. A hybrid machine learning model is proposed that leverages unsupervised and supervised algorithms to classify network traffic as normal or malicious with high accuracy. The CICDDoS2019 dataset containing real and simulated attack traffic is used to train and evaluate various classification algorithms. Feature selection is applied to extract the most relevant attributes for detection. The proposed ensemble model combines K-Means clustering, random forest, extreme gradient boosting, adaptive boosting, support vector machine, and artificial neural network classifiers. Experimental results demonstrate the model achieves 99.65% overall accuracy in distinguishing DDoS attacks, outperforming individual algorithms. This research contributes towards developing more robust techniques for analyzing network behaviors and enhancing intrusion detection capabilities.

Keywords: DDoS · K-Means · machine learning · ensemble modeling · CICDDoS2019 dataset · network security

1 Introduction

With the growing reliance on internet-based services, distributed denial of service (DDoS) attacks pose a serious threat due to their ability to disrupt network accessibility and functionality. DDoS attacks overwhelm victim systems and deplete critical resources like CPU, memory, and bandwidth through a flood of malicious traffic (Yuan et al., 2017). Attackers leverage compromised devices to launch coordinated assaults, amplifying their impact. Distinguishing malicious traffic from legitimate user activity is challenging, especially as attack vectors evolve rapidly. Early and accurate detection is crucial for mitigating damage [1]. This research aims to develop an effective DDoS detection system leveraging machine learning techniques. A hybrid approach is proposed that combines unsupervised and supervised algorithms to classify traffic patterns. The contributions of this work are:

(1) A novel ensemble model integrating K-Means clustering, random forest, XGBoost, AdaBoost, SVM and artificial neural network classifiers.

A. Bagwari et al. (Eds.): ICEWTS 2024, CCIS 2399, pp. 95–106, 2025.
https://doi.org/10.1007/978-3-031-87886-2_7

(2) Feature selection using chi-square test to extract the most relevant attributes from a large network traffic dataset.
(3) Evaluation of the proposed model on the CICDDoS2019 benchmark dataset containing real and simulated attack traffic.
(4) Comparative analysis demonstrating improved detection accuracy over individual algorithms.

The remainder of the paper is organized as follows: Sect. 2 discusses related work on DDoS detection methods. Section 3 describes the CICDDoS2019 dataset and pre-processing steps. Section 4 presents the proposed methodology. Section 5 reports experimental results and performance comparisons. Finally, conclusions are drawn in Sect. 6.

2 Related Work

Several machine learning based techniques have been proposed for DDoS attack detection which can be broadly categorized as anomaly-based, signature-based, and hybrid approaches (Deepa et al., 2018) [2]. Anomaly-based techniques establish a baseline of normal traffic behaviour and flag deviations as potential attacks (Wankhede and Kshirsagar, 2018). However, they suffer from high false alarm rates due to normal traffic variations. Signature-based methods match traffic patterns to predefined attack signatures but cannot detect novel attacks (Ye et al., 2018). Hybrid methods leverage the strengths of both by employing anomaly detection to identify unknown threats along with signature matching for known attacks, providing improved coverage (Deepa et al., 2018) [3]. Ensemble models combining multiple algorithms have also shown promise through their complementary strengths [4]. Researchers have applied various machine learning algorithms for classification including random forest, decision trees, naive bayes, k-nearest neighbors, support vector machines, neural networks and deep learning models [5]. Feature selection techniques help reduce dimensionality and improve performance (Abdulhammed et al., 2018). Studies have evaluated algorithms on benchmark datasets like CICIDS2017 and CICDDoS2019 (Aytac et al., 2020b; Maranhao et al., 2020; Sindian & Sindian, 2020). While prior work has explored individual algorithms, this research proposes a novel hybrid ensemble approach combining unsupervised K-Means clustering with supervised learning classifiers for improved DDoS attack detection (Table 1).

3 Dataset

The CICDDoS2019 dataset [21] from the Canadian Institute for Cybersecurity is used, containing network traffic records with labeled normal and attack instances (Fig. 1).

Key characteristics of the dataset:

- 88 network traffic attributes like source/destination IP, port, packet lengths, and flow durations
- Over 4.9 million records comprising normal traffic and 13 types of DDoS attacks
- Attacks include SYN flood, ICMP flood, HTTP flood, UDP flood, DNS amplification

Suitable for training, testing and evaluating machine learning models.

Table 1. Systematic review of DDoS traffic attack detection previous works

Paper	Dataset	Classifier(s)	Accuracy
Abdulhammed et. al. [6]	AWID	RF, AdaBoost, MLP J48, Logit Boost	99.64%,
Belouch et. al. [7]	UNSW- NB15	SVM, RF, NB, DT	97.49%
Filho et. al. [8]	CICIDS2017, CICIDS2018	RF, AdaBoost, LR, SGD, DT	96.0%
Chen et. al. [9]	CICDDoS2017, KDDCUP99	MCNN	99.18%
Elejla et. al. [10]	Generated Flow-based Datasets	MLP, DT, NB, SVM, KNN	85.0%
Hosseini et. al. [11]	NSL-KDD	NB, RF, DT, MLP, KNN	98.9%
Aytac et. al. [12]	CICDDoS2019	ANN, SVM, LR, KNN, DT, RF, NB	99.7%
Maranhao et. al. [13]	CICDDoS2019, NSL-KDD	ADA, LDA, LR, RF	98.78%
Sindian et. al. [14]	CICDDoS2019	DNN using Autoencoder, EDSA	96.3%
Polat et. al. [15]	Self Generated	SVM, NB, ANN, KNN	98.3%
Elsayed et. al. [16]	CICDDoS2019	RNN-Autoencoder	99.0%
Salahuddin et. al. [17]	CICDDoS2019	Autoencoders	99.0%
Cil et. al. [19]	CICDDoS2019	FNN (128 units, 3 hidden layes)	99.9%
Shieh et. al. [20]	CICDDoS2017, CICDDoS2019	Bidirectional LSTM, GMM	%

3.1 Data Collection and Features

Table 2 shows the file sizes and number of records for each traffic type in the dataset. The dataset utilized in this study encompasses details regarding 88 network traffic attributes and event logs, offering insights into 13 distinct types of DDoS attacks, as displayed in Table 2.

4 Proposed Methodology

DDoS attacks present a substantial menace to network infrastructure, potentially causing service disruptions by inundating target systems with malicious traffic [22]. A potent strategy to counter these risks revolves around constructing a DDoS attack detection system that melds unsupervised and supervised machine learning methodologies [23]. This hybrid approach harnesses the robust attributes of techniques to promptly discern and

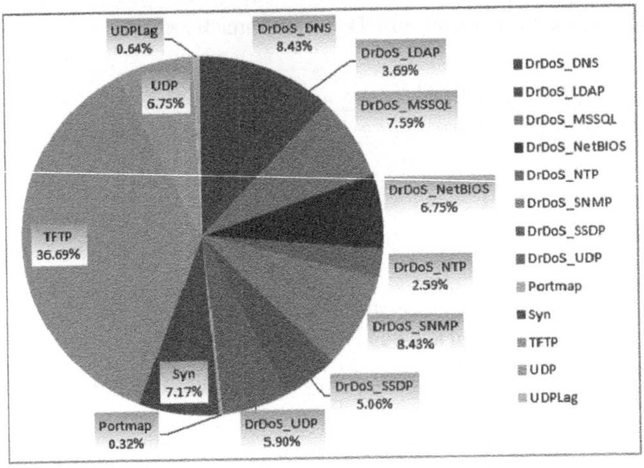

Fig. 1. Various network traffics in the CICDDoS2019

Table 2. File size and records in the CICDDoS2019

S. No.	File Name	File Size
1	DrDoS_DNS	2 GiB
2	DrDoS_LDAP	874 MB
3	DrDoS_MSSQL	1.8 GiB
4	DrDoS_NetBIOS	1.6 GiB
5	DrDoS_NTP	615 MB
6	DrDoS_SNMP	2 GiB
7	DrDoS_SSDP	1.2 GiB
8	DrDoS_UDP	1.4 GiB
9	Portmap	74.9 MB
10	Syn	1.7 GiB
11	TFTP	8.7 GiB
12	UDP	1.6 GiB
13	UDPLag	150.7 MB

thwart these assaults, thereby reinforcing network security and safeguarding the availability and dependability of services. Figure 2 illustrates the proposed hybrid machine learning methodology for DDoS attack detection consisting of the following phases:

Phase 1: Data Preparation.

The CICDDoS2019 dataset is acquired and subjected to pre-processing as described earlier to clean, normalize and partition the data.

Phase 2: Algorithm and Hyperparameter Selection.

Relevant classification algorithms are identified based on literature review. Their hyperparameters are optimized using grid search.

Phase 3: Building the Classification Model.

The model construction involves:

(1) Applying K-Means clustering on training data to generate pseudo-labels.
(2) Training individual classifiers on randomly distributed labelled data.
(3) Combining classifiers in an ensemble model for prediction.

Phase 4: Model Evaluation.

The trained ensemble model is tested on reserved data and performance metrics like accuracy, precision, recall etc. are computed.

The methodology leverages unsupervised clustering for dataset extension followed by supervised ensemble learning for classification. Feature selection and hyperparameter tuning aid model optimization.

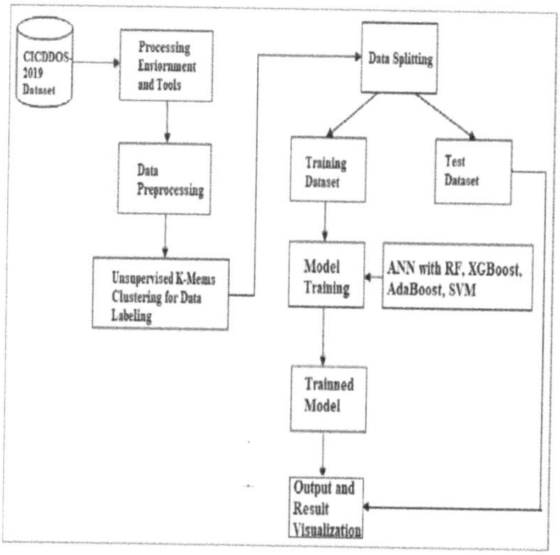

Fig. 2. Proposed Methodology for DDoS Attack Detection

PHASE 1: DATA PREPARATION.

i. Handling Missing Values

This holds significance as the absence of data points can substantially influence the precision of machine learning models [24]. Managing missing data is carried out through the application of suitable strategies, including imputation methodologies (mean, median, or mode), or the elimination of rows featuring missing values. The dataset underwent an initial scrutiny for missing data utilizing Python's 'isnull()' method.

ii. Check for duplicates

The dataset was subject to initial inspection for duplicate entries using the duplicate method in Python. Any instances displaying identical sets of features in two or more traffic flows were excised from the dataset. Subsequent verification ascertained the complete removal of duplicates. The absence of duplicate entries within the CICDDoS2019 dataset negates the need for duplicate data management during the data pre-processing phase.

iii. Removing Outliers from Dataset

Outliers, typically denoting data points that deviate markedly from the dataset's distribution patterns or structure, pose a challenge in data processing. The identification and removal of these outliers necessitate a systematic assessment of data dispersion and the pinpointing of data points that significantly stray from the dataset's norms. Outliers can impede machine learning algorithms, hindering their generalization and resulting in inaccurate outputs. After numerous iterations and a thorough analysis of threshold values, we ascertained an optimized threshold value of 2 for accurate outlier identification in our Python script, illustration shown in Fig. 3. Sample output of outliers of few features is shown here:

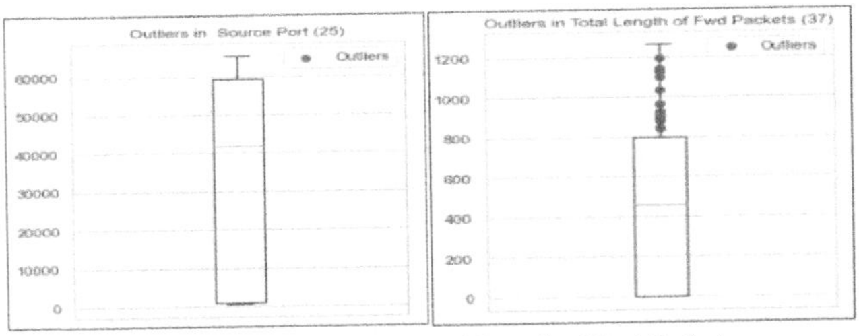

(a) Source Port (b) Total Length of Fwd Packets

Fig. 3. Outliers found in different features

iv. Feature Selection

The feature selection process was applied to dataset during the pre-processing phase. This involves selecting a subset of the most important features for use in machine learning models. This is crucial because it can improve the accuracy of machine learning models while reducing the computational complexity of training and inference. The goal is to identify a subset of features that are highly relevant for distinguishing between normal and attack traffic. In this model, the Chi-square feature selection technique was used after cleaning the data to identify the top features for detecting DDoS attacks. This process results in a set of 15 highly relevant features using the 'SelectKBest()' Python function.

PHASE 2: SELECTION OF ALGORITHMS AND THEIR HYPERPARAMETERS.

This phase explains how the most appropriate algorithms for detecting DDoS attacks were selected.

A. **Algorithm Selection Criteria.**

From the literature review Table 1, most suitable algorithms have selected for our proposed approach. The selection of algorithm is based on four criteria that we defined in Table 3. These criteria made it possible to develop this research and fulfill the proposed objective.

Table 3. Criteria for classification Algorithm Selection

S. No.	Selection Criteria
01	Captures complex patterns, sequential and non-linear relationships
02	Handles mixed data types
03	Improves weak learner algorithms
04	Effective for multi-class classification
05	The algorithm is available in the Python library
06	Low computational complexity

B. **Optimized Hyper-parameters used in Algorithms.**

In this proposed methodology, parameters for the classifiers are listed here in Table 4. Here are the best tuned parameters for the proposed hybrid machine learning models along with their corresponding values:

PHASE 3: BUILDING THE CLASSIFICATION MODEL.

Step 1: Apply K-means Algorithm
The K-means unsupervised machine learning algorithm has been employed on the dataset. This choice stems from the necessity to enable real-time classification of incoming data packets using machine learning algorithms. In real-time scenarios, procuring labelled data for incoming packets is often unfeasible. They facilitate the creation of a dataset extension, wherein new auto-generated labels are assigned to the data.

Step 2: Dataset Division
While the literature commonly advocates a 70% training and 30% testing data split, specific conditions may necessitate alternate distribution strategies. In our research, a higher emphasis was placed on training. Consequently, the dataset was randomly partitioned into two subsets: training set (comprising 80% of the data) and a test set (constituting 20% of the data).

Step 3: Random Distribution of Training Data
The concept of randomizing the allocation of training data across diverse machine learning classification algorithms is a process designed to enhance the overall classification task performance through a ensemble approach.

Step 4: Training by Ensemble Learning
Ensemble learning is a machine learning paradigm wherein multiple models or algorithms are combined to make predictions. The objective is to amplify overall performance by harnessing the complementary strengths of individual models. In our proposed

Table 4. Hyper-parameters used in selected algorithms

Model	Parameters	Optimized Values
Artificial Neural Network	Hidden Layer Sizes	(100, 100)
	Activation Function	relu
	Solver	adam
	Regularization Parameter (Alpha)	0.001
Random Forest	Minimum Samples Split	2
	Maximum Depth	None
	Number of Estimators	150
	Minimum Samples Leaf	1
XGBoost	Number of Estimators	100
	Learning Rate	0.1
	Maximum Depth	6
	Subsample	0.8
AdaBoost	Base Estimator (Decision Tree)	Max. Depth: 1
	Number of Estimators	150
	Learning Rate	0.1
SVM	Kernel	rbf
	Gamma	0.01
	C	10
	Random state	42

model, illustrated in Fig. 4, we employ five supervised machine learning classification algorithms to create an ensemble model. These algorithms include Artificial Neural Network, Random Forest, XGBoost, AdaBoost, and SVM.

Step 5: Performance Comparison
Following the evaluation of the trained model, the outcomes were systematically organized and subjected to comparative analysis, using predefined reference variables. This comparison facilitated the assessment of the proposed model's effectiveness and enabled us to implement any required refinements to achieve the highest efficiency. The results of each trained model are documented in Table 1.6. If the resulting criterion is not met, we proceed to fine-tune the algorithm parameters. This iterative process is continued until the performance of the proposed model.

- Improved Data Segmentation: K-Means clustering groups data into distinct clusters, allowing ensemble classifiers to tailor their learning approach within more homogeneous subsets, improving accuracy.
- Enhanced Model Robustness: Ensembles trained on clusters reduce individual classifier biases, leading to more stable, reliable predictions.

- Noise Reduction: Clustering helps filter out noise by isolating outliers, enabling classifiers to generalize better and handle complex data patterns.

5 Experiment Results and Model Performance Comparison

5.1 Results

i. **Feature Selection**

The chi-square test was applied to rank features based on their correlation with the target variable and select the most relevant subset. 15 top-ranked attributes were retained which captured important traffic characteristics while reducing dimensionality. The performance summary of all modules with algorithm's accuracy according to features is shown in Table 5.

Table 5. Module wise attack detection accuracy

No. of Features	Module-1		Module-2		Module-3		Module-4	
	ANN	RF	ANN	XGBoost	ANN	AdaBoost	ANN	SVM
5	0.9527	0.9638	0.9589	0.9764	0.9616	0.9768	0.9598	0.9689
10	0.9718	0.9767	0.9731	0.9819	0.9759	0.9841	0.9794	0.9765
15	0.9935	0.9967	0.9931	0.9914	0.9859	0.9971	0.9894	0.9978
20	0.9907	0.9914	0.9925	0.9897	0.9894	0.9967	0.9818	0.9834
25	0.9826	0.9829	0.9786	0.9761	0.9795	0.9718	0.9804	0.9738

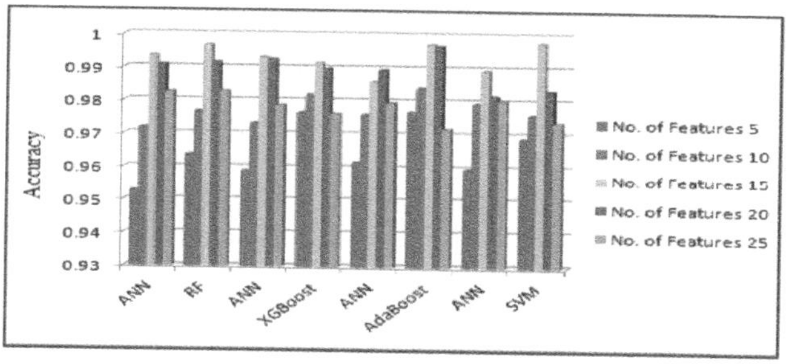

Fig. 4. Attack detection accuracy with different features

ii. **Performance Comparison**

The pre-processed data with selected features served as input to train the proposed ensemble model. Individual classifiers - random forest, XGBoost, AdaBoost, SVM and neural network were first trained on randomly distributed labeled data generated via K-Means clustering. Their predictions were then combined through majority voting. The accuracy detection illustrate in Fig. 4. The trained model was evaluated on reserved test data and classification accuracy along with other metrics was computed as shown in Table 2. The experiment began with 25 features, followed by the extraction of 20, 15, 10, and 5 features through the chi-square method. In this study, we explored the feature sets within the range of 5–25 features to identify the highest accuracy while minimizing the number of features. To optimize the accuracy, parameters were fine-tuned for the proposed classification model, as indicated in Table 6.

Table 6. Attack detection accuracy of various ensemble modules

No. of Features	Ensemble ANN and RF	Ensemble ANN and XGBoost	Ensemble ANN and AdaBoost	Ensemble ANN and SVM
5	0.9582	0.9676	0.9692	0.9643
10	0.9742	0.9775	0.9800	0.9777
15	0.9951	0.9922	0.9965	0.9936
20	0.9910	0.9911	0.9930	0.9826
25	0.9827	0.9773	0.9756	0.9771

As seen, the ensemble model achieved 99.65% overall accuracy. This validates the hypothesis that combining multiple algorithms leverages their complementary strengths for improved detection performance. A comparative analysis was also conducted by training the classifiers on datasets with varying number of features selected using chi-square. As shown in Fig. 5, the ensemble model achieved highest accuracy of 99.46% using the optimal 15 features. This demonstrates that feature selection aids model optimization by retaining only the most discriminative attributes necessary for classification while reducing dimensionality and complexity. The proposed ensemble approach based on unsupervised clustering and supervised learning exhibited state-of-the-art results for DDoS attack detection on the CICDDoS2019 benchmark, validating its effectiveness. The model can generalize well to detect both known and emerging threats (see Fig. 5).

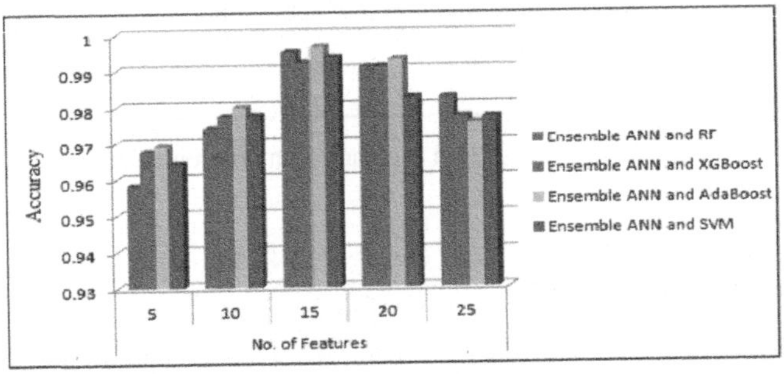

Fig. 5. Performance comparisons of various ensemble modules

6 Conclusion

This research addressed the critical problem of detecting distributed denial of service attacks through a novel hybrid machine learning methodology. A model was developed leveraging unsupervised K-Means clustering and an ensemble of supervised classifiers including random forest, XGBoost, AdaBoost, SVM and neural network. Experiments were conducted on the large-scale CICDDoS2019 network traffic dataset containing real and simulated attacks. Feature selection helped reduce dimensionality while retaining important traffic characteristics. The proposed ensemble classification approach achieved 99.46% overall accuracy in distinguishing DDoS attacks, outperforming individual algorithms. This research contributes towards developing more robust techniques for analyzing network behaviors and enhancing intrusion detection capabilities. Future work will focus on improving model scalability, integrating real-time traffic analysis capabilities, exploring behavioral patterns to detect zero-day threats, and building user-friendly tools to strengthen network defense mechanisms against evolving cyber-attacks.

References

1. Yuan, X., Li, C., Li, X.: DeepDefense: Identifying DDoS Attack via Deep Learning. 2017 IEEE International Conference on Smart Computing (2017). https://doi.org/10.1109/smartcomp.2017.7946998
2. Wankhede, S., Kshirsagar, D.: DoS attack detection using machine learning and neural network, 2018 Fourth International Conference on Computing Communication Control and Automation, Pune, India, pp. 1–5, IEEE (2018). https://doi.org/10.1109/iccubea.2018.8697702
3. Hasan, M.Z., Hasan, K.M.Z., Sattar, A.: Burst header packet flood detection in optical burst switching network using deep learning model. Proc. Comput. Sci., **143**, 970–977 (2018). https://doi.org/10.1016/j.procs.2018.10.337
4. Ye, J., Xiang-Yang, C., Zhu, J., Feng, L., Song, L.: A DDOS attack detection method based on SVM in software defined network. Secur. Commun. Netw. **2018**, 1–8 (2018). https://doi.org/10.1155/2018/9804061
5. Deepa, V., Sudar, K.M., Deepalakshmi, P.: Detection of DDoS attack on SDN control plane using hybrid machine learning techniques. 2018 International Conference on Smart Systems and Inventive Technology (2018). https://doi.org/10.1109/icssit.2018.8748836
6. Abdulhammed, R., Faezipour, M., Abuzneid, A., Alessa, A.: Effective features selection and machine learning classifiers for improved wireless intrusion detection. 2018 International Symposium on Networks, Computers and Communications (2018). https://doi.org/10.1109/isncc.2018.8530969
7. Belouch, M., El Hadaj, S., Idhammad, M.: Performance evaluation of intrusion detection based on machine learning using Apache Spark. Proc. Comput. Sci., **127**, 1–6 (2018). https://doi.org/10.1016/j.procs.2018.01.091
8. De Lima Filho, F.S., Silveira, F.A.F., De Medeiros Brito Júnior, A., Vargas-Solar, G., Silveira, L.F.Q.: Smart detection: an online approach for DoS/DDoS attack detection using machine learning. Secur. Commun. Netw., **2019**, 1–15 (2019). https://doi.org/10.1155/2019/1574749.
9. Chen, J., Yang, Y., Hu, K., Zheng, H., Wang, Z.: DAD-MCNN. Proceedings of the 2019 11th International Conference on Machine Learning and Computing 19 (2019). https://doi.org/10.1145/3318299.3318329

10. Elejla, O.E., Belaton, B., Anbar, M., Alabsi, B., Al-Ani, A.K.: Comparison of Classification Algorithms on ICMPv6-Based DDoS Attacks Detection. Computational Science and Technology. Lecture Notes in Electrical Engineering, vol 481. Springer, Singapore (2019). https://doi.org/10.1007/978-981-13-2622-6_34
11. Hosseini, S., Azizi, M.: The hybrid technique for DDoS detection with supervised learning algorithms. Comput. Netw. (2019). https://doi.org/10.1016/j.comnet.2019.04.027
12. Aytac, T., Aydin, M.A., Zaim, A.H.: Detection DDOS attacks using machine learning methods. Electrica **20**(2), 159–167 (2020). https://doi.org/10.5152/electrica.2020.20049
13. Maranhao, J.P.A., Costa, J.P., De Freitas, E.P., Javidi, E., De Júnior, R.T.S.: Error-robust distributed denial of service attack detection based on an average common feature extraction technique. Sensors **20**(20), 5845 (2020). https://doi.org/10.3390/s20205845
14. Sindian, S., Sindian, S.: An enhanced deep autoencoder-based approach for DDOS attack detection. WSEAS Trans. Syst. Control, **15**, 716–724 (2020). https://doi.org/10.37394/23203.2020.15.72
15. Polat, H., Polat, O., Cetin, A.E.: Detecting DDOS attacks in software-defined networks through feature selection methods and machine learning models. **12**(3), 1035 (2020). https://doi.org/10.3390/su120310035
16. Elsayed, M. S., Le-Khac, N.-A., Dev, S., Jurcut, A.D.: DDoSNet: a deep-learning model for detecting network attacks. 2020 IEEE 21st International Symposium on "A World of Wireless, Mobile and Multimedia Networks" (WoWMoM) (2020). https://doi.org/10.1109/wowmom49955.2020.000
17. Salahuddin, M.A., Faizul Bari, M., Alameddine, H.A., Pourahmadi, V., Boutaba, R.: Time-based Anomaly Detection using Autoencoder. 2020 16th International Conference on Network and Service Management (2020). https://doi.org/10.23919/cnsm50824.2020.926911
18. Gaur, V., Kumar, R.: Analysis of machine learning classifiers for early detection of DDOS attacks on IoT devices. Arab. J. Sci. Eng., 1353–1374 (2021). https://doi.org/10.1007/s13369-021-05947-3
19. Cil, A.E., Yildiz, K., Buldu, A.: Detection of DDoS attacks with feed forward based deep neural network model. Expert Syst. Appl. **169**, 114520 (2021). https://doi.org/10.1016/j.eswa.2020.114520
20. Shieh, C., Lin, W., Nguyen, T., Chen, C., Horng, M., Miu, D.: Detection of unknown DDoS attacks with deep learning and Gaussian mixture model. Appl. Sci. **11**(11), 5213 (2021). https://doi.org/10.3390/app11115213
21. Halladay, J., Cullen, D., Briner, N., Warren, J., Fye, K., Basnet, R. B., Bergen, J., Doleck, T.: Detection and characterization of DDOS attacks using Time-Based features. IEEE Access, **10**, 49794–49807 (2022).https://doi.org/10.1109/access.2022.3173319
22. Fouladi, R.F., Ermis, O., Anarim, E.: A Novel Approach for distributed denial of service defense using continuous wavelet transform and convolutional neural network for software-Defined network. Comput. Secur. **112**, 102524 (2022). https://doi.org/10.1016/j.cose.2021.102524
23. Rani, S.J., et al.: Detection of DDoS attacks in D2D communications using machine learning approach. Comput. Commun. **198**, 32–51 (2023). https://doi.org/10.1016/j.comcom.2022.11.013
24. Rajput, D.S., Upadhyay, A.K.: Hybrid technique for DDOS attack detection using machine learning (2022). www.philstat.org. https://doi.org/10.17762/msea.v71i4.1403

Design and Performance Evaluation of an Energy-Efficient 16x16 SRAM Memory Array for Low Power Applications

P. Anuradha[1](✉), Ch. Navitha[1], B. Khaleelu Rehman[1], and G. Renuka[2]

[1] Department of ECE, Chaitanya Bharathi Institute of Technology, Hyderabad, Telangana, India
anuradhap_ece@cbit.ac.in

[2] Department of ECE, Anurag University, Hyderabad, Telangana, India

Abstract. Memory plays a vital role in any semiconductor device. There are two types of memory: SRAM (static random access memory) and DRAM (dynamic random access memory). Major design limitations include small functional space, power consumption, and latency to design any electronic component. Power consumption is one such factor which needs optimization in Memory. This article is related to designing of 16X16 memory array using 6T SRAM. The optimization was done in the sense amplifier in the memory architecture where the power consumption of the circuit is optimized from 447 micro watts to 350 micro watts. The delay of the circuit is also optimized from 225 ns to 133 ns. The design is implemented in cadence virtuoso tool.

Keywords: Charge pump · transient · layout · power gated · phase noise · pow6T SRAM · Sense amplifier · Cadence Virtuoso

1 Introduction

In today's electronic landscape, the fusion of creative memory cells are important for storing data retrieval or backup. Data stored in memory cells which are the fundamental components of any semiconductor device. To satisfy today's advance technology advancements we need memory cell which are basic block of semiconductors to have low power consumption, less leakage current, high performance, cost efficient. Since memory cells are essential for both data storage and retrieval, they are the fundamental components of any semiconductor device. The 0s and 1s are stored in these memories. Depending on what we require, we can read, write, and erase the binary data. The structure of memory in computer memory systems is defined by memory architecture. A computer system's memory architecture determines its capacity and performance. Memory cells fall into a variety of kinds, including Flash Memory, Dynamic Random Access Memory, and Static Random Access Memory (SRAM). As a result, we designed a 16x16 SRAM array for low power consumption and less delay which are mostly used for various areas like embedded systems, biomedical devices and military etc.

A. Bagwari et al. (Eds.): ICEWTS 2024, CCIS 2399, pp. 107–117, 2025.
https://doi.org/10.1007/978-3-031-87886-2_8

2 Related Work

In this era of constant development of SRAM (Static Random Access Memory) design is motivated by the search for memory solutions that are both power consumption efficient and performance effective. Many approaches to reduce leakage power, boost stability, and increase performance in SRAM cells have been researched recently.Shalini Singh presents a 7T SRAM cell-based 1KB memory array with reduced leakage and latency, while Jayram Shrivas presents a unique 7T SRAM cell that uses a sleep technique and high-k gate dielectric material [1]. [2] Rashmi Bisht places a strong theory on low power consumption and noise reduction, while [3] Himanshu Banga shows how to build an area efficient SRAM with less leakage. Noise-free SRAM cells with significantly improved amplitude, speed and leakage current static performance were also introduced by Shyam Akashe. These advance development highlights the search for effective and more advance SRAM solutions in the face of changing technological requirements. [4] Jesal P. Gajjar shows how to build 32-bit SRAM array using CMOS technology. It compares the parameters such as power and area by implementing the design in 90 nm and 45 nm technology.

Aruru Sai Kumar shows how to implement 32 × 32 memory array (1KB) and analysed parameters such as power consumption, leakage power, delays etc. They made comparative analysis by changing basic cell such as 7T and 8T [5].Many innovative solutions have been presented by different authors: an array of 1 Kbyte memory that reduces leaks and latency. An architecture that prioritizes read stability and optimizes billing. Noise reduction methods; Effective territorial design; SRAM 7T cells with sleep technique and high dielectric gate material; New SRAM cells with improved size, speed and leakage characteristics. These techniques emphasize the ongoing efforts to develop reliable and efficient SRAM solutions that meet changing technology needs, with an emphasis on finding improvements and energy-efficient memory solutions [6] [7].

3 Proposed Method

Memory array was designed by 7T and 8T SRAM cells, implemented the memory array using 6T SRAM cell which would consume less power. The memory architecture consists of 6T SRAM cell, Pre-charge circuit, Sense amplifier, Writer Driver circuit and Decoder. Our proposed design has sense amplifier with capacitor instead of pass transistors [8]. This would help in not draining the complete vdd on bit or bit bar lines thus reducing the power consumed by the circuit [9].

A. *SRAM Architecture.*

The Framework of SRAM architecture is depicted in Fig. 1. It consists of row and column decoder, Writer driver circuit, Pre-charge circuit, 6T SRAM cell, Sense amplifier [10]. A SRAM cell can store 1 bit i.e. binary digit. As we are implementing 16 × 16 memory array there would be total of 256 SRAM cells capable of storing 256 bits of data [9], [10]. So, 4×16 decoders are utilized as row and column decoders. As we are using 6T SRAM cells the power consumption is reduced.

B. *Sense Amplifier.*

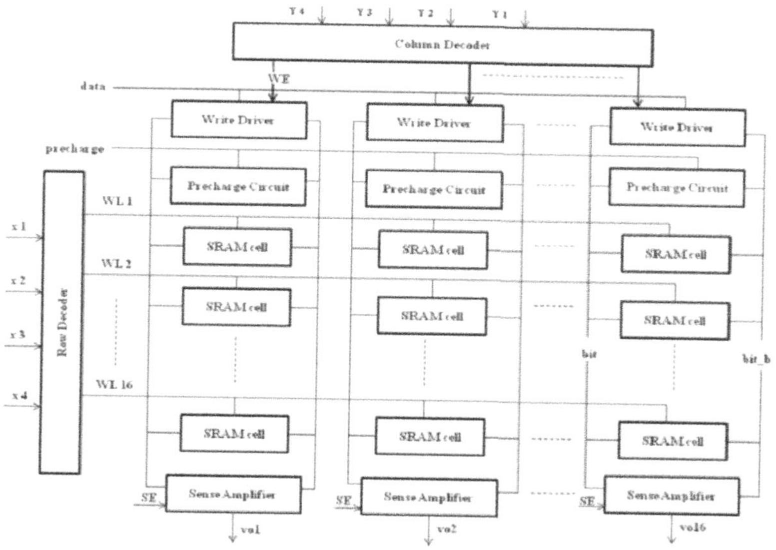

Fig. 1. Block Diagram of SRAM memory array construction

Sense amplifier is utilized to amplify the small differential voltages in the circuit. The Sense amplifier circuit is shown in Fig. 2. SRAM cells required are smaller.

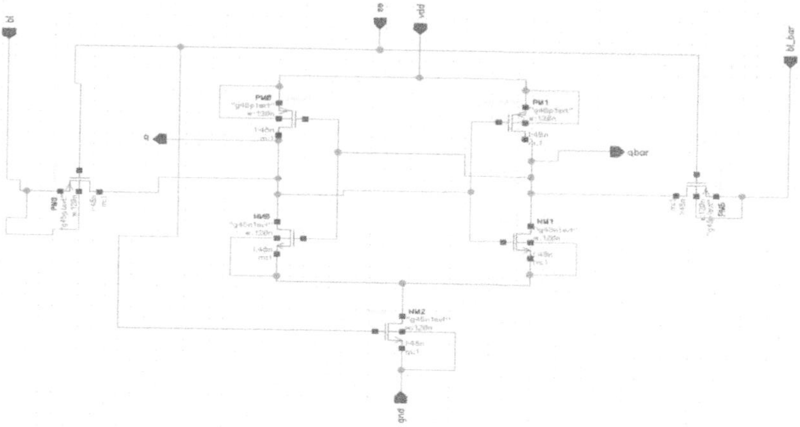

Fig. 2. Sense Amplifier with Transistors

when sense amplifier is used since there is no need of bit lines to discharge completely [11], [12].

C. *Modified Sense Amplifier.*

The two PMOS transistors acting as pass transistors in Fig. 3 are replaced with capacitors of capacitance 100 pf. This modified sense amplifier consumes less power than because two transistors are removed and there is no need of complete draining of bit lines from vdd. Modified circuit is shown in Fig. 3.

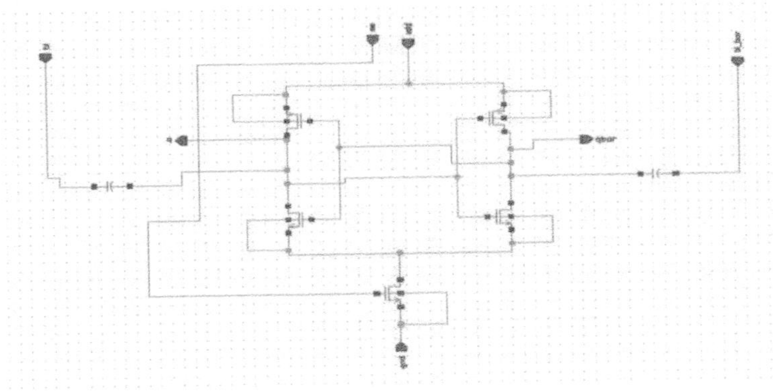

Fig. 3. Sense Amplifier with Captivators

D. *Pre Charge Circuit.*

It is used to charge both bit and bit bar lines to VDD. This is required during the read operation [13, 14]. The pre-charge circuit is shown in Fig. 4. Each bit line ends with PMOS coupled to the precharge circuit. This circuit uses an equalizer transistors to input the bit current to the vdd and make them perfectly balanced when charging the battery before setting to 0.

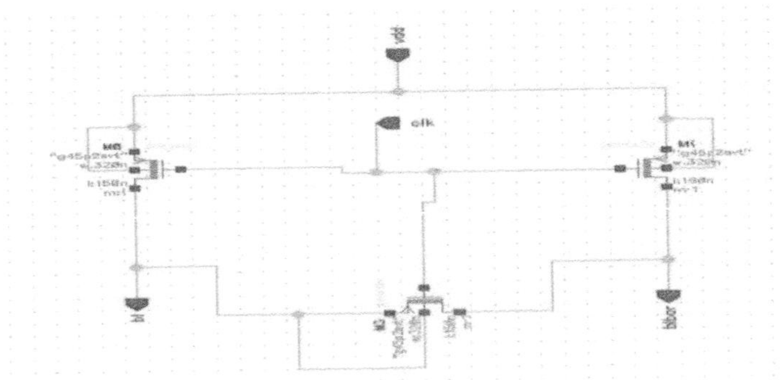

Fig. 4. Precharge Circuit

E. *Writer Driver Circuit.*

It is used to write data on to the SRAM cells during write operation. The circuit consists of 3 inverters and 4 PMOS transistors. The circuit is shown in Fig. 5.

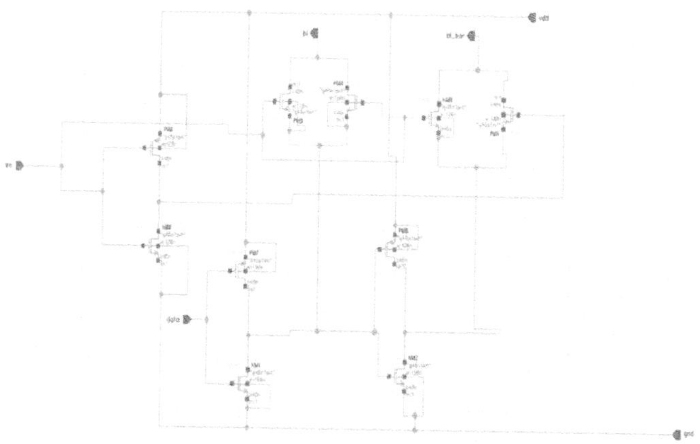

Fig. 5. Writer Driver Circuit

4 Result and Analysis

Test bench to check the functionality of the circuit is consists of a dc source with voltage VDD for transistors to work. Then it has ac source called Vpulse which is given as input to the address lines, data, sense enable and clk. Then we have gnd to connect the lower terminals of all the sources and block to the ground. Output is observed at terminals 0–15. The schematic consists of 2 decoders, pre-charge circuit, sense amplifier, 6-T SRAM cell and writer driver circuit. The output waveform of precharge circuit is shown in Fig. 6 Net 1 describes vdd power supply.Net 2 describes the clock. q1 describes bit line and q2 describes bit bar line. As we can see that at every clock cycle both bit and bit bar lines are charged to 1.5 V i.e. vdd from 0.25 and 0.75 V. This wave forms clearly depicts the working of pre charge circuit where the both bit and bit bar lines are charged up to vdd which is very much important during read operation. The output waveform of sense amplifier circuit is shown in Fig. 7 where the output voltages present on bit and bit bar lines is amplified. Net 1 indicates the sense enable. Net 2 indicates vdd. Net 4 indicates input bit line.Net 3 indicates input bit bar line. q Indicates the amplified output of bit line and q_bar indicates the amplified output of bit bar line. As we can see that the pairs qout, bit line and qout bar, bit-bar line is same but the qout and qout bar voltages are amplified from 1.1 V to 2.27 V and 0.1 to 1.76 V.

The output waveform of writer driver circuit is shown in Fig. 8 where the data given as input is copied on to both bit and bit bar lines. Net 3 indicates write enable. Net 2 indicates data. Net 1 indicates vdd power supply, q1 indicates output of the writer driver circuit which is given as input to SRAM cell during write operation, and q2 indicates complemented output which is given as input to bit bar line during write operation. During write enable (net3) high the data (net 2) is written on to bit (q1) line. The same data is been complemented on to the bit bar line (q2).

6T SRAM cells perform two basic operations: reading and writing. During the read process, the bit and bit lines are treated as output lines, the stored data is transferred from the cell to the bit and bit lines. In contrast, bit and bit lines serve as input paths

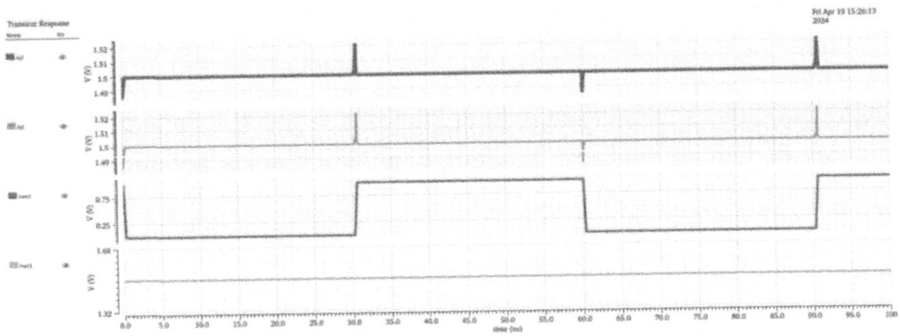

Fig. 6: Precharge Circuit Output Waveform

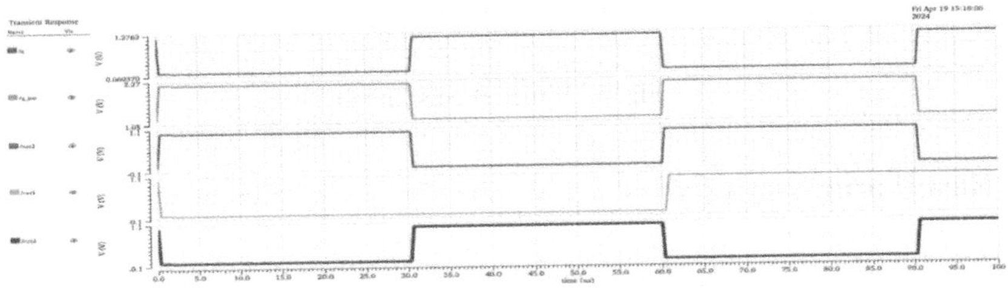

Fig. 7. Sense Amplifier Outputs

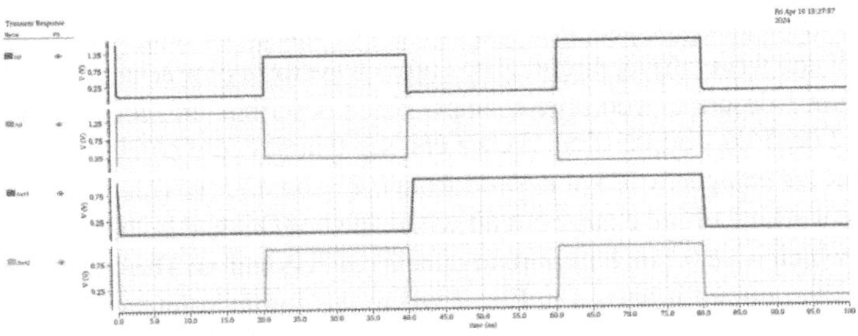

Fig. 8. Writer Driver Outputs

during the write process. The data to be written is first placed on the bit and bit lines, and then it is transferred to the cell. As Alan Turing said: "The machine surprises me with great frequency," and in this case the efficiency of the SRAM cell always does. Figure 11 describes the resulting waveform during the write operation. The output waveform of 6T SRAM cell is shown in Fig. 9 where the data written on bit and bit bar lines is copied in to the cell during write operation and vice versa during read operation. Net 5 indicates that the word vdd.

During the read process, bit and bit lines act as output lines. The bit and bit lines are charged to vdd by the pre-input circuit. When the enable word is high, the input transistor is turned on and the data on the panel is transferred to the bit and bit lines. As we can see, if net 6 (Writable) is high, qout is net4 (bit line) and qout bar is net 8 (bit bar).

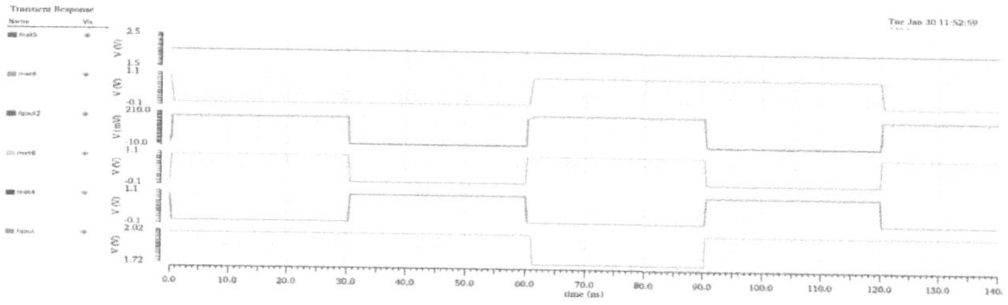

Fig. 9. 6T SRAM Cell Outputs

Figure 10 describes the learning activity. As we can see, net4 (bit-string) qout and net 8 (bit-string) are the same as output band q as long as net 6 (Writable) is high. Figure 11 describes output waveform of 16X16 Memory architecture. The below waveforms are only the inputs given to the memory unit such as address lines, data.

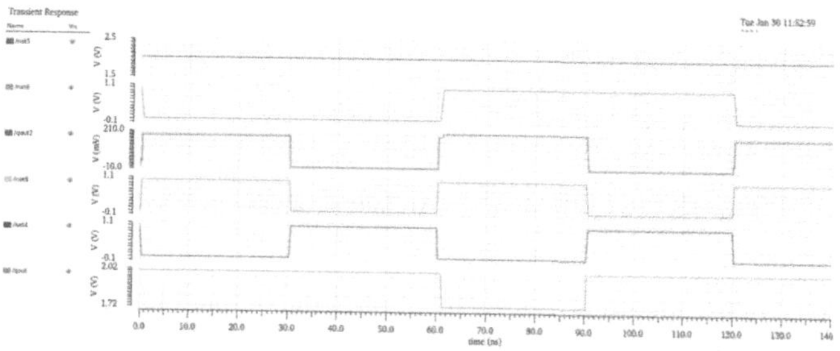

Fig. 10. 6T SRAM Cell Output Waveform during read operation

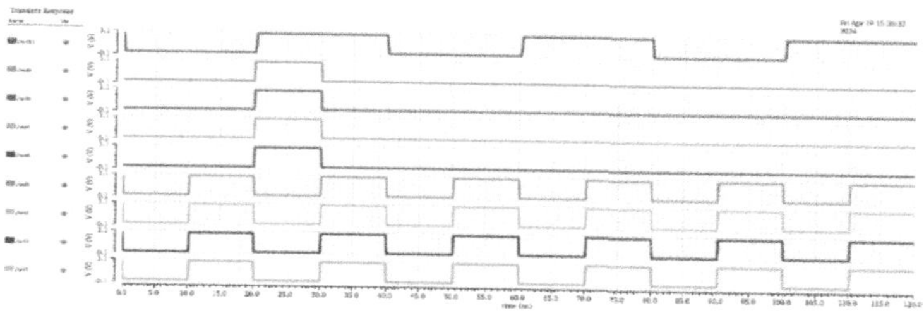

Fig. 11. 16 × 16 Memory Architecture Output Waveform

Figure 12 describes inputs such as clk, sense enable and outputs from 0–2. Net 2 indicates row address line from row decoder ar. Net 3 indicates row address line 2 from row decoder br. Net 4 indicates row address line 3 from row decoder cr. Net 5 indicates row address line 4 from row decoder dr.Net 6 indicates column address line 1 from column decoder ac. Figure 13 describes remaining outputs from 0–9.

Figure 14 describes remaining outputs from 7–15. As we can observe that the data (net 11) is copied in to the cells (0–9). The above describes the data been copied into

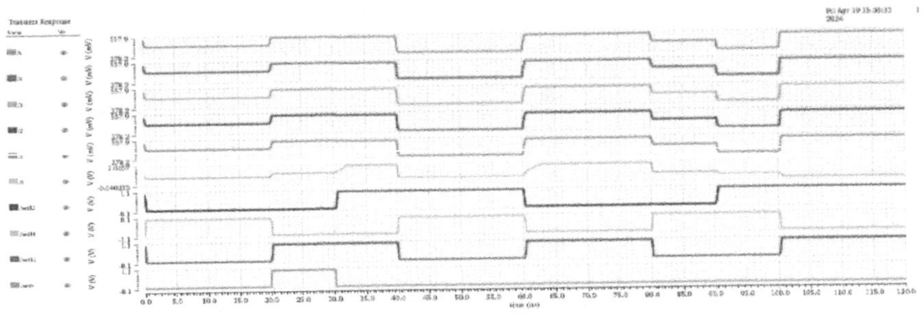

Fig. 12. 16 × 16 Memory Architecture Output Waveform

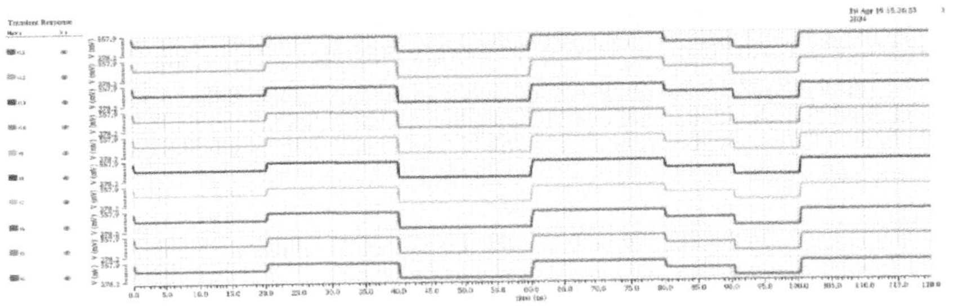

Fig. 13. 16 × 16 Memory Architecture Output Waveform

the output cells. Sense amplifier is used to amplify the output obtained at bit and bit-bar lines. Sense Amplifier output waveforms is given in Fig. 15.Net 1 indicates the sense enable.Net 2 indicates vdd. Net 4 indicates input bit line. Net 3 indicates input bit bar line. q indicates the amplified output of bit line. q_bar indicates the amplified output of bit bar line.

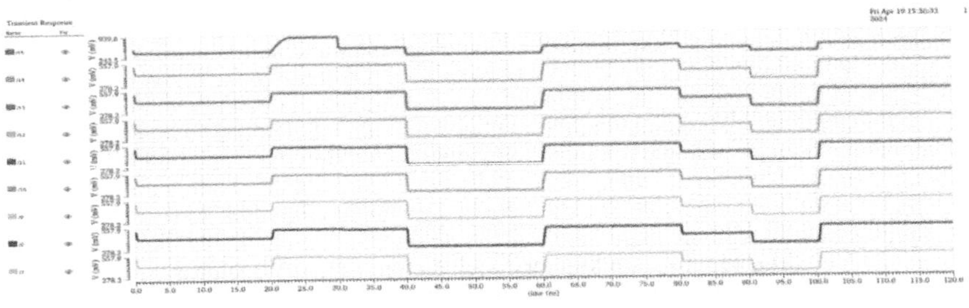

Fig. 14. 16x16 Memory Architecture Output Waveform

Figure 16 describes output waveform of 16X16 Memory architecture. The below waveforms are only the inputs given to the memory unit such as address lines, data.

Figure 17 describes inputs such as clk, sense enable and outputs from 0–2. Net 2 indicates row address line from row decoder ar. Net 3 indicates row address line 2 from row decoder br. Net 4 indicates row address line 3 from row decoder cr. Net 5 indicates

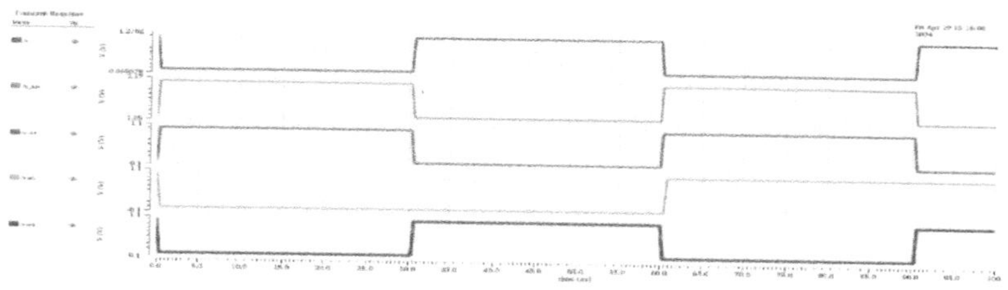

Fig. 15. Sense Amplifier with Capacitor Output Waveform

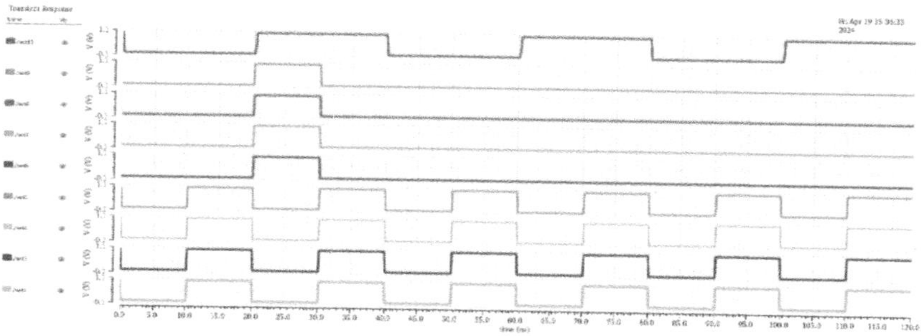

Fig. 16. 16 × 16 Memory Architecture with Capacitor Output Waveform

row address line 4 from row decoder dr.Net 6 indicates column address line 1 from column decoder ac. Figure 18 describes remaining outputs from 0–9. As we can observe that the data (net 11) is copied in to the cells (0–9). The clock signal pre charges the bit and bit bar lines and the data read will be transferred from bit and bit bar lines on to the sense amplifier. When the sense enable is high the same data is retrieved in the cells i.e. high value of output is retained during time 30ns to 50 ns.

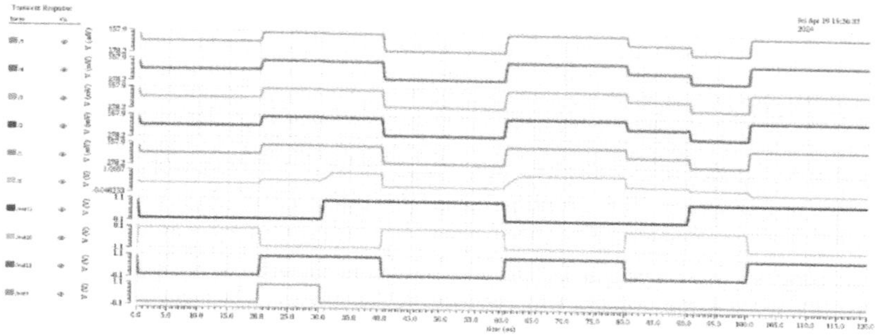

Fig. 17. 16 × 16 Memory Architecture with Capacitor Output Waveform

Figure 19 describes remaining outputs from 7–15. As we can observe that the data (net 11) is copied in to the cells (0–9). The above describes the data been copied into the output cells. Power consumption and delay comparative analysis of 16x16 Memory Architecture with Capacitor and transistor is shown in Table 1.

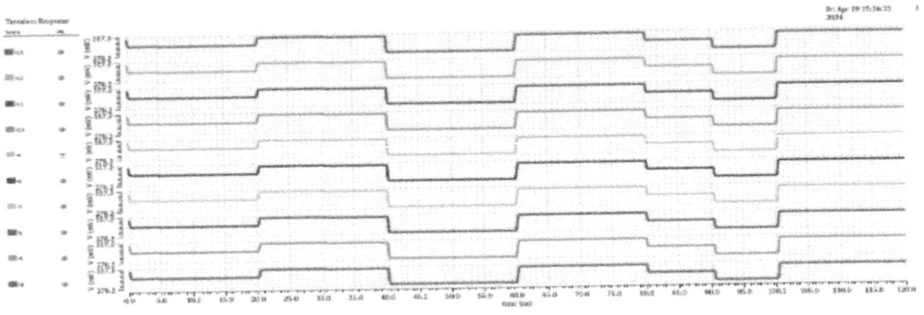

Fig. 18. 16 × 16 Memory Architecture with Capacitor Output Waveform

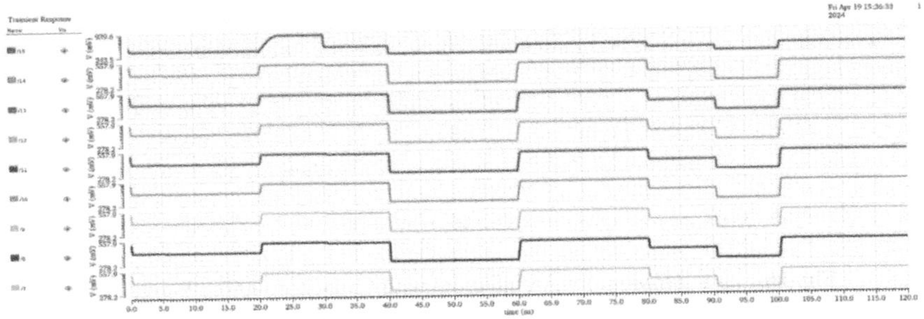

Fig. 19. 16x16 Memory Architecture with Capacitor Output Waveform

Table 1. Comparison of power consumption.

Quantity	With Transistor	With Capacitor
Power Consumption (uW)	447	350
Delay(ns)	225	133

5 Conclusion

By using a sense amplifier with five transistors instead of seven, our approach further minimizes power usage. The coupling capacitors are used in place of the two missing transistors. Unlike transistors, which require the voltage to be completely drained in order to observe the output at the sense amplifier, capacitors do not require this. In light of this, the circuit is using little electricity. However, there is a trade-off between area and power because capacitors take significantly more space than transistors. This memory architecture works well in low power applications where power, not area, is the primary design constraint. Thus, the aforementioned memory architecture can be used in mobile devices, wearable health technology, aeronautical systems, and military equipment, among other things.

References

1. Singh, S., Akashe, S.: Low-power 1 KB (32 × 32) compact array memory using 7T SRAM cells. Wire. Pers. Commune. **96**, 1099–1109 (2017)
2. Bisht, R., Aggarwal, P., Karki, Ph.D., Pande, P.: Low Power and Noise Resistant 16 × 16 SRAM Array Design using CMOS Logic and Differential Sense Amplifier. In the process of the 2016 International Conference on Computer Communications and Automation (ICCCA), Greater Noida, India, pp. 1474–1478 (29–30 April 2016)
3. Banga, H., Agarwal, D.: Implementing a 16 × 16 SRAM memory array. In the run-up to the 2017 International Conference on Innovation in Signal Processing and Embedded Systems (RISE), Bhopal, India, pp. 458–462 (October 27–29, 2017)
4. Gajjar, J.P., Zala, A.S., Aggarwal, S.K.: Design and analysis of 32-bit SRAM architecture in 90 nm CMOS technology. In Proceedings of International Research Journal of Engineering and Technology, Pune, India (2016)
5. Xue, C., Kumar, A.S., Khalaf, O.I., Sominini, R.P., Abdulsahib, G.M., Sujith, A., Danuja, T., Vinay, M.V.S.: Design and performance analysis of 32 × 32 Memory Array SRAM for low power applications. In the collection of 2022 articles in microelectronics
6. Gadepalli, B.V., Somineni, R.P., Peddy, A., Gianniekode, U.M.: Design and analysis of 16nm GNRFET and CMOS 4kb SRAM low power array based on 1-bit 6T SRAM cell. In continuation of the IEEE IAS Global Conference on Emerging Technologies (GlobConET), Arad, Romania (20–22 May, 2022)
7. SVS Prabhu Deva Kumar, A., Shaival Suman, B., Overview, C.: Stability and performance analysis of low power 6T SRAM cells and memristor SRAM cells using 45NM CMOS technology. In the process of the 2018 International Conference on Innovation in Electrical, Electronic and Communication Engineering (ICRIEECE) Bhubaneswar, India (27–28 July, 2018)
8. Kalesvari, B., Kaja Mohidin, S.: Design, implementation and analysis of 8T SRAM cells in memory arrays. Int. Ch. Eng. Technol., **7**, 101–105 (2018); Renuka, G., Anuradha, P., Reddy, P.L., Gurumurthy, V., Rajkumar, K.: Implementation of TCAM controller enabled CDMA network on chip router for high speed 5G communication. SN Inf., **4** (6), 740. 2023
9. Anuradha, P., Navitha, S., Renuka, G., Jitender Reddy, M., Rajkumar, K.: An in-depth research framework developed by Harris Hawkes algorithm for intelligent ECG classification in WSN-IoT. Environment. J. Intell. Fuzzy Syst., (Preprint), pp. 1–13 2023
10. Srivastava, P., Chauhan, R.C.S.: Design of power efficiency and voltage phase sensors for PLL applications in 45nm CMOS technology. J. Shanghai Univ. Sci. Technol. **23**(11), 184 (2021)
11. Samdani, J., Anuradha, P., Rao, H.: Architecture of high-speed polar encoders using VLSI LDPC codecs. J. Adv. Res. Dyn. Control Syst. **10**(10), 153–161 (2018)
12. Kumar, Y., Raman, A., Ranjan, R., Deep, S., Sarin, R.K.: Low power RVCO differential, low phase noise and wide tuning range for PLL applications. Int. J. Electron. **111**(2), 379–392 (2024)
13. Anuradha, P., Rajkumar, K., Arabelli, R., Shareena, R.: Implementation of high speed 64-bit MAC unit using FPGA. In AIP Conference Proceedings, Vol. 2418, No. 1, p. 030060. AIP Publishing LLC (May 2022)
14. Anuradha, P., Rajkumar, K., Arabelli, R.R., Shareena, R.: QCA-based error detection circuit for nanocommunication network. In Smart Intelligent Computing and Applications, Volume 1: Proceedings of Fifth International Conference on Smart Computing and Informatics (SCI 2021), pp. 141–149. Singapore: Springer Nature Singapore (2022, April)

Author Index

A. Bagwari et al. (Eds.): ICEWTS 2024, CCIS 2399, p. 119, 2025.
https://doi.org/10.1007/978-3-031-87886-2